UNCERTAINTY
BY DESIGN

EXPERTISE

**CULTURES AND
TECHNOLOGIES
OF KNOWLEDGE**

EDITED BY DOMINIC BOYER

A list of titles in this series is available at cornellpress.cornell.edu.

UNCERTAINTY BY DESIGN

Preparing for the Future with Scenario Technology

Limor Samimian-Darash

CORNELL UNIVERSITY PRESS   ITHACA AND LONDON

Sections of chapter 1 appeared previously in "Governing Uncertainty, Producing Subjectivity: From Mode I to Mode II scenarios," *Subjectivity* 14, no. 6 (2021): 1–18, co-authored with Michael Rabi.

Sections of chapter 2 and chapter 3 appeared previously in "Practicing Uncertainty: Scenario-Based Preparedness Exercises in Israel," *Cultural Anthropology* 31, no. 3 (2016): 359–86.

First published 2022 by Cornell University Press

Library of Congress Cataloging-in-Publication Data

Names: Samimian-Darash, Limor, author.
Title: Uncertainty by design : preparing for the future with scenario technology / Limor Samimian-Darash.
Description: Ithaca [New York] : Cornell University Press, 2022. |
    Series: Expertise: cultures and technologies of knowledge |
    Includes bibliographical references and index.
Identifiers: LCCN 2021026441 (print) | LCCN 2021026442 (ebook) |
    ISBN 9781501762451 (hardcover) | ISBN 9781501762468 (paperback) |
    ISBN 9781501762475 (epub) | ISBN 9781501762482 (pdf)
Subjects: LCSH: Uncertainty—Social aspects. | Strategic planning—Simulation methods—Social aspects. | Decision making—Simulation methods—Social aspects. | Strategic planning—Case studies. | Decision making—Case studies.
Classification: LCC HM1101 .S255 2022 (print) | LCC HM1101 (ebook) |
    DDC 003/.5—dc23
LC record available at https://lccn.loc.gov/2021026441
LC ebook record available at https://lccn.loc.gov/2021026442

For Chosro and Shulamit

Contents

Preface

In my graduate research (at HUJI), I examined preparedness for nonconventional biological threats in Israel, and the unique form that such preparations took in that country. I discovered an assemblage of multiple experts and units involved in conceptualizing biological "exceptional" events as well as diverse ways to prepare for them, which I termed a "pre-event configuration." Although I came to see the distinct nature of different future events and the heterogeneity dynamism of the state-preparedness formation, I did not explicitly address "uncertainty" as a concept until after the completion of that writing project.

A subsequent reflection on the two cases of a smallpox vaccination project and preparedness for pandemic flu launched my journey toward the concept of uncertainty. The smallpox vaccination project conducted in Israel in the winter of 2002–2003 during the Second Gulf War reflected preparedness for the possibility of a future smallpox event in which both the biological agent and the vaccine against it were already known. Preparing for pandemic influenza, however, invoked a different problematic. The particular virus strain that might be involved in a future outbreak was unknown, and thus the disease's clinical case definition could not be ascertained in advance. That is to say, until an influenza pandemic takes place, it is impossible to know what specific viral agent will be involved and how exactly to prepare for it. In such a situation, however, an epidemic is neither a fabrication nor an abstraction; rather, the potential for its appearance already exists, and this virtual occurrence can actualize as different events in the future—as various pandemic strains that may require different types of treatment and response.

It was then that I realized that a smallpox pandemic and pandemic flu each represented a different type of uncertain future. Over time, I came to refer to these two types of uncertainty as possible uncertainty and potential uncertainty. The concept of uncertainty was not a subject of particular interest within anthropology at the time. Nor was the future as dominant a site of research and critique as it is today. It was only when I visited as a postdoctoral fellow at the University of Illinois at Urbana-Champaign, through the generous hospitality of Virginia Dominguez and Jane Desmond, that I decided it was time to study and reflect on the concept of uncertainty itself—first by reading everything that had been written on the topic in a range of disciplines and second by developing a conceptualization based on my fieldwork research, which I thought could open up an interesting discussion for anthropologists.

My first attempt to publish the theoretical and ethnographic piece in which I set out this conceptualization, however, was unsuccessful. The first journal to which I submitted my article did not even send it out for review, claiming, ironically, that the topic was not sufficiently global (or central) within anthropology. I was about to give up on publishing the article, thinking that I had taken my conceptualization of uncertainty and related theories too far, but instead I took the advice of a colleague who suggested I send it elsewhere. The second attempt, with *Current Anthropology*, was not just different; it was a diametrically opposed experience. In a short period, the article was published in full, along with an additional five commentaries, and opened up a whole new intellectual conversation on the topic, both for me and for others, in this field.

In the following two years of postdoctoral research at Stanford University (hosted by James Ferguson), I concluded what proved to be a very exciting period by coediting a book with Paul Rabinow titled *Modes of Uncertainty: Anthropological Cases* and published by the University of Chicago Press. Though an edited volume, this was not an ordinary collection of articles. Rather, it was an experiment. We asked the contributors to take our initial framework on risk and uncertainty and to test it in their fieldwork in different domains, such as finance and markets, security and humanitarianism, and health and the environment.

Until that time, I had been viewing uncertainty as an ontological problem, examining how different governmental technologies sought to respond to that problem. Only later, when focusing on emergency preparedness exercises in Israel, did I begin to think of it in a different way.

Uncertainty was not just a future event or threat to prepare for. Rather, it appeared to be part of the rationality of the preparedness technology itself. Scenario-based exercises, I came to see, were involved with the notion of uncertainty in a different way—exemplifying what I now regard as an uncertainty-based technology. Although the scenario is based on a preselected, well-designed event, once practiced it is actualized into multiple incidents that the various participants enact which have unexpected consequences. Through the use of this technology, participants are directed neither toward predicting the future nor toward discovering the best solutions for an unknown future. Instead, the technology generates uncertainty through its execution, from which new problems are identified.

After conducting ethnographic research on Israel's annual Turning Point preparedness exercises, I proposed to go beyond the field of security and to examine scenarios as a management form rooted in uncertainty-based thinking in the fields of health and energy and how they are used in thinking of and practicing global future problems. In addition, I took the scenario technology as my primary object of research, bringing the strengths of ethnographic research directly to bear on it. This time, I looked at scenarios at the World Health Organization and the World Energy Council. I assumed at the outset that the scenario technology would be expressed differently in these two organizations in relation to the distinctive problems facing their respective fields of activity. In each case, I examined how scenarios were designed to meet the organizations' specific conceptions of the future and its uncertainties.

Scenario experts commonly approach the scenario technology from a normative position and sometimes see it as a tool to provide more freedom for our societies. My critical engagement is slightly different. I do not inquire into the efficiency or effectivity of scenarios as a means of addressing future uncertainty. Rather, I am interested in how the scenario became a predominant technology in contemporary society and how new modes of veridiction, jurisdiction, and subjectivation are expressed through this technology. I am interested in how uncertainty-based technologies, and scenarios in particular, not only promise to open up future possibilities but also design uncertainties through their practice.

As with any philosophical effort, once concepts are extracted from actualities—that is, counter-effectuated—they are then put into readers' hands and minds to test, to reactualize in new sites, as a way of identifying new technologies and new worlds of meaning and action. I therefore hope that the discussion of uncertainty and the scenario technology presented in this work will stimulate further explorations of those topics within the field of anthropology and beyond.

Let me end with a quotation from *The Logic of Sense*, by Gilles Deleuze ([1969] 1990, 56–57): "This problem does not at all express a subjective uncertainty, but, on the contrary, it expresses the objective equilibrium of a mind situated in front of the horizon of what happens or appears." This book summarizes a long journey from initial thoughts on uncertainty to a fully developed analysis of uncertainty-based technologies and scenarios together with an examination of the topic in multiple cases. It is not a journey intended to represent, deal with, or solve subjective (the experience of) uncertainty. Rather, it is a journey to open up the mind, to move from notions of knowledge, control, possibility, and assessment toward an objective equilibrium in which uncertainty is seen as a site of potentiality reflected in the ways in which knowledge, practices, and subjectivity are experienced.

This book is the fruit of a half-a-decade-long journey in which I had the opportunity to discuss my research projects with many friends and colleagues. I have been lucky to have these people in my life, and I hope I have adequately expressed my deep appreciation to them over the years.

I am deeply thankful to Paul Rabinow for years of mentoring, inspiring conversations, and sincere collaboration. From the first days of my initial visit to UC Berkeley as a student until my return as a visiting professor, Paul had always been a challenging voice, an open mind, and a generous heart. Much of my academic experience would have been very different without his guidance and support. Paul specifically helped me to start writing this book, and his influence might be discerned in many of its pages.

I am especially grateful to Eyal Ben-Ari, who has been both a mentor and a colleague, guiding me in my academic career over the years. Eyal has played a huge part in my accomplishing this book. In his modest way, not seeking to take any credit, he read every chapter of the manuscript and provided crucial suggestions. He is a great example of what support and pure giving mean.

I am grateful to Don Handelman, Don Brenneis, and James Faubion—each of whom dedicated hours of conversations and helped in developing many of the ideas that were eventually incorporated into this work. Their long support, mentoring, and generosity cannot be summed up in a few words. I cherish them very much. I cannot forget that it was Don Handelman who first introduced me to the work of Deleuze and Guattari. Every meeting with him was full of intellectual excitement, and many ideas from our conversations have ultimately found their way into this book.

I am especially grateful to the Van Leer Institute in Jerusalem, which hosted me for a sabbatical year between 2018 and 2019. Hagai Boaz and Shay Lavi provided me with the right atmosphere and a conducive intellectual space in which to write this book.

I acknowledge funding from the Israel Science Foundation for two research grants, during the years 2015–2018 and 2019–2022 (grant numbers 1635/15 and 1120/19, respectively). This support enabled a long-term fieldwork study on Israel's national preparedness exercises, my subsequent study of the World Energy Council, and my inquiry into the World Health Organization's exercises. The grants also allowed me to hire and support outstanding students and research assistants. I would like to thank Michael Rabi for his invaluable work throughout the past few years. More than a student or a research assistant, Michael has been a colleague with whom I enjoyed collaborating on many projects. Specifically, Michael made a huge contribution to what became chapter 1 and the epilogue of the present work, which are the result of many conversations and our analysis of the history of scenario planning. I also thank two research assistants, Tzofia Goldberg and Nir Rotem, who were part of the earlier stages of the studies on which this book is based.

I am particularly indebted to the people whom I interviewed and talked to during my research for this book. These communications included long-term fieldwork, official interviews, and many informal talks. This book is the result of their generous willingness to share with me parts of their professional life and experience.

Understanding the professional side of scenario planning and experts would not have been possible without the generous help from and fascinating conversations with Gerald Davis and Angela Wilkinson, two world leaders in the field. I cherish their patient acceptance of my anthropological questions, observations, and interventions.

The book as a whole, as well as each individual chapter, has benefited from suggestions from numerous colleagues from a range of disciplines. I specifically thank (in alphabetical order) Gaymon Bennett, Irit Dekel, Stefan Elbe, Hedva Eyal, Turo-Kimmo Lehtonen, Yagil Levy, Avi Shoshana, and Meg Stalcup. I thank my colleagues and friends in the School of Public Policy at the Hebrew University of Jerusalem for their long-standing support. I am also thankful for the many opportunities I was granted to present my work and receive helpful commentary on it. I am specifically grateful to the 2018–2019 Polonsky Academy Fellows of the Van Leer Institute, Jerusalem; the 2019 Workshop on Enacting Uncertainties at the University of Helsinki and my hosts Turo-Kimmo Lehtonen and Janne Hukkinen; and the 2019 International Workshop on Scenarios and the Politics of the Future at the Climate–Security Nexus, University of Hamburg, and my hosts Susanne Krasmann and Christine Hentschel.

In the process of bringing this manuscript into publication, I was fortunate to come into contact with Jim Lance, the senior editor for anthropology and social sciences at Cornell University Press. Working with Jim has made the journey of publishing this book a very positive, instructive, and strengthening experience. I am grateful for the opportunity he gave me to publish with Cornell and for the professional guidance he provided along the way.

John Carville has worked with me on the copyediting of the manuscript from its early stages. His precise and excellent work has helped me better put my ideas into written form.

Above all, I thank my husband, Asaf Darash, with whom I have shared more years together than apart. He has been everything I could wish for and has given me unlimited support, kindness, and love. He and our two sons, Erel and Yahav, have been a constant source of happiness and reason throughout the many paths I have taken in my life.

In Commemoration of Paul Rabinow

In the days when I was reviewing the final edits of this book, Paul Rabinow passed away. "Tell me something interesting," he would say at the beginning of a meeting. And he would truly be interested in what the person in front of him would say, regardless of whether they were young students or established scholars.

My meetings with Paul would end up with him helping me organize my ideas, providing new challenges, and charging me with more power to continue my intellectual journey. For so many years, he had been there for me. From when I was a young anonymous student arriving from Israel for the first time through to when I returned as a visiting professor, he was always there—a great mind, with an even greater heart.

Everyone knows that Paul was an extraordinary intellectual, thinker, and philosopher. But he was also a great teacher and educator. Through his daily interactions with students, his guidance, and his collaboration, he nurtured an entire generation of academics. Paul left a spark of life and passion within each and every one of his students. The flame that he helped kindle within me will continue to burn in the hope that I will also be able to pass it on to my own students.

Thank you, Paul.

Rest in peace.

Abbreviations

EIS	event information site
IHR	international health regulations
JADE	joint assessment and detection of events
NEMA	National Emergency Management Authority
NFP	National Focal Point
PHEIC	public health emergency of international concern
WHO	World Health Organization
WHO	Europe World Health Organization's Regional Office for Europe

UNCERTAINTY BY DESIGN

INTRODUCTION
Uncertainty, Scenarios, and the Future

In creating scenarios, we are like explorers setting out on a journey of discovery. Although we have a vision of our destination, we know that conditions may change, new opportunities—or problems—will probably emerge, and the chances are we will need to change direction and adjust our course. Building and using scenarios helps us to face the rigours of our journey. It prepares us to maintain an open mind and be flexible in the face of uncertainty.

—Ged Davis

Humankind has long struggled with the uncertainty of the future and particularly with how to foresee the future, imagine alternatives, or prepare for and guard against undesirable eventualities. This book does not provide answers to such problems. It is neither a manifesto that sets out to describe an undesirable state of affairs that might be termed the "uncertainty society" nor a toolkit for designing the types of futures we might wish to see. Instead, my aim is to problematize different ways in which societies conceptualize and act on the uncertain future and to understand how and why particular social technologies have emerged accordingly as well as how these technologies reshape and affect the ways in which we experience our world. My principal objective within this effort is to shed light on one particular technology for systematically thinking, envisioning, and preparing for future uncertainties: *the scenario.*

The scenario—or scenario planning—emerged to become a widespread means through which states, large corporations, and local organizations imagine and prepare for the future. Regardless of the different ways in which scenarios are created and used, various iterations of the scenario technology all share a common approach to thinking and practicing potential futures: they narrate imagined stories about the future that are intended to help people move beyond their "mental blocks" to consider the "unthinkable" (Bradfield et al. 2005; Ramírez and Wilkinson 2016; Ringland 1998; Schwartz 1996). What I find most striking about the scenario technology, though, is how it differs from other means usually associated with managing future uncertainties (i.e., calculations and evaluations based

1

on knowledge of past events, leading to possible prediction, control, or prevention of unknowable futures), which draw mainly on the construct of risk and the related notion of risk management. Instead of risk, I argue that the scenario draws on the construct of uncertainty and promotes a particular approach to the governing of the future that I term *uncertainty by design*. As scenario-planning expert Ged Davis (2002) puts it, scenario planning is a "journey of discovery" that helps us to remain open to "new opportunities—or problems—[that] will probably emerge," fostering a degree of flexibility "in the face of uncertainty." In the scenario, in other words, embracing the unpredictability of the future is a central aspect of the methods used to address—indeed, to describe and prepare for—that future.

Scenarios create stories of the future neither to translate the future into assessed possibilities nor to predict it in advance but rather to identify new potentials, to destabilize perceptions of the future as closed or certain, and to mitigate overreliance on existing knowledge and models in efforts to address the unknown future. Put differently, the scenario technology not only accepts the potential uncertainty of the future but also promotes uncertainty as a mode of observing and acting in the world.

The term *uncertainty by design*, it should be noted, is not meant to imply that the scenario technology is based on an assumption that the future can be molded into a specific and determined image. In using the term *design*, I mean that the scenario technology embraces uncertainty rather than reduces it, embarks on the unknown future instead of attempting to predict it. What is signified by this term is an approach that accepts the open-endedness of the future while at the same time facilitating techniques to manage it that go beyond practices based on or involving the calculation of possibilities. *Uncertainty by design* signifies a mode of governing based on imagination, potentiality, and acceptance of the emergent and the unpredictable. Yet it is a *designed* practice, one that has specific rules and systems for creating, narrating, and using the future stories that make up the scenario. In addition, participants in scenario planning experience uncertainty in a particular way. They are not just (external) users of something previously created through the technology; instead, they themselves are involved in the creation and practicing of scenario narratives. Through this designed process, participants learn, embody, and create new ways of understanding, approaching, and experiencing uncertainties. Hence, the scenario technology takes the condition of uncertainty as one of its key premises, and uncertainty is expressed in the technology's modes of veridiction (knowledge making), jurisdiction (power, exercising), and subjectivation (the experience of the subject) (Foucault 1991; Rabinow and Bennett 2012), as will be discussed below.

I first began to think about uncertainty as a phenomenon and concept while conducting research on the governance of future threats in health and security.[1]

It was in this context that I first deployed the concept of potential uncertainty (Samimian-Darash 2013, 2016). Drawing on Deleuze and Guattari's (1987) concepts of the virtual and the actual, I argue that potential uncertainty derives from the variety of actualities that can emerge from a virtual event rather than from lack of assessments of future possibilities. Observing the future through the construct of risk assumes both that it is possible and that one has the capacity to calculate and evaluate future possibilities on the basis of knowledge of past events. Observation through the construct of uncertainty, however, is built on the assumption that the future and its unfolding into a broad variety of events (potential rather than "possible") cannot be known in advance, calculated, or assessed.

Uncertainty is thus fundamentally distinct from risk on both ontological and epistemological grounds. Accordingly, the technologies based on the two concepts are also dissimilar. If risk-based technologies manifest the constraints and limitations of future unknowns, transforming them into possibilities (even if there are multiple possibilities) and making and marking probabilities, uncertainty-based technologies do not "make us free" (Bernstein 1998; O'Malley 2004) but rather provide a distinct form of governing the future through the acceptance and proliferation of situations of potentiality and unpredictability. Future unknowns, then, can be managed or governed in different ways, with risk-based techniques being but one possible figure in a broad future-governance problematization.[2]

In the analysis set out in this book, I avoid taking a merely ontological approach to the issue of risk and uncertainty in the world—an approach that is usually accompanied by a metatheoretical framework that seeks to explain the difference between premodern and modern societies.[3] At the same time, my analysis seeks to go beyond the sociocultural epistemological approach to differences in risk constructions within different cultures, which often remains at a representational level.[4] While I acknowledge the importance of such approaches for analyzing and theorizing cultural differences and societal historical shifts, my interest in uncertainty is driven by the governmental approach, which provides a new conceptual framework for understanding both ontological and epistemological notions related to uncertainty as well as associated technologies. Addressing this anthropologically means to "inquire into what is taking place without deducing it beforehand" (Rabinow 2007, 3; see also Rabinow 2003)—in other words, to study uncertainties, the governmental technologies that seek to address them, and related human behavior, with an analytical approach that makes it possible to uncover the diverse conceptualizations of the future that are involved in these phenomena and the complex formations in concrete forms of life to which they give rise (Samimian-Darash 2013). At the object level, in ontological terms, the questions I ask include what is uncertainty, what are events or situations of uncertainty in the world, and are there different types of uncertainty or uncertain futures. I propose using the terms *potential uncertainty* and *possible uncertainty* to better differentiate between different types of unknown futures. At the conceptual epistemological level, I differentiate between different conceptions of that which is perceived as a future threat or unknown—that is, *danger, risk*, and *uncertainty*—where each of these terms refers to a distinct perception of the reason or source of unknown events. Then, at the governmental level, I ask how different future problematizations emerge and how rendering the future through different conceptions (e.g., risk or uncertainty) affects the emergence of governing technologies that express specific rationalities and determine particular types of solutions.

My analytical effort involves shedding light on technologies based on the rationalities of uncertainty that have evolved as ways of governing the unknown future—what I term *uncertainty-based technologies* (Samimian-Darash 2016). While existing scholarship has paid some attention to governmental forms of risk-based technologies such as insurance (Dean 1999; Ewald 1991; Lobo-Guerrero 2011), urban risk management (Zeiderman 2016), geriatric assessment (Kaufman 1994), community psychiatry (Rose 1996), and genetic counseling interviews (Rapp 1995), the study of uncertainty-based technologies (Samimian-Darash 2016) remains lacking at both the empirical and the conceptual levels. Addressing this lacuna, then, I examine scenario planning as an uncertainty-based technology and identify its particular mode of governing as that of uncertainty by design.

Anthropological Cases

Historically speaking, the scenario technology emerged within the fields of security and civil defense (Lakoff 2007) and has subsequently diffused into many other domains, in the process becoming a central technology in efforts to prepare for the future in today's world. Strikingly, scenarios are used today within many different fields of knowledge and expertise, including in the contexts of climate change, security, health, and energy (see, for example, Cooper 2010; Faubion 2019; Lakoff 2008; Samimian-Darash 2016). Scenarios are used to envisage, think through, and better prepare for threats as well as to open up new opportunities and ways of thinking in relation to potential future progress and innovation.

Although scenarios are currently being deployed in many spheres of human activity, they have rarely been comprehensively studied from a sociocultural perspective. Most social scientific studies to date have inquired into scenarios only secondarily in the course of examining other broad topics such as security and health preparedness or broader theoretical frameworks such as anticipation and resilience and have largely relied on documentary sources.[5] Within these pages, however, I take the scenario as my primary object of study, bringing the strengths of anthropological research to bear on it directly.[6]

I present cases of the use of scenario technology in the fields of security and emergency preparedness, energy, and health by analyzing scenario narratives and practices at the National Emergency Management Authority (NEMA) in Israel, the World Health Organization's Regional Office for Europe (WHO Europe), and the World Energy Council. These three different spaces enable comparative analysis along a number of different axes. The World Energy Council is a registered charity, a pragmatic and neutral global member-based network; WHO Europe represents an international organization operating at the regional level; while Israel's NEMA is a state-level public organization with advisory, policy, and operative functions. All three organizations produce and practice scenarios on different scales and with different spreads (i.e., the extent to which the scenarios they enact have implications beyond their respective organizational or locational boundaries and are concerned with multiple future trends and threats). The three different fields examined also diverge in terms of their temporal orientations and how the past, present, and future interrelate in the creation of scenario scripts. With their distinctive configurations and distinctive fields of knowledge, expertise, and temporality, the three sites thus together provide a novel research base for understanding scenario thinking and its contemporary spread in today's world.

Following a discussion of the origin and development of the scenario technology, I first present my findings from a long-term, fieldwork-based research project on scenario-based emergency exercises in Israel. These exercises, known as Turning Point (Hebrew: *Nekudat Miffne*), build on nationwide scenarios and are intended to prepare all governmental ministries, local municipalities, emergency operation units, and the general population for future emergencies (e.g., war or earthquake). This particular case allowed me to observe both the ways in which scenarios were built and narrated over months of preparations and the way in which they were subsequently put into practice and exercised. That is to say, I followed the creation of a large-scale national scenario narrative and then observed the participation of various organizations in the exercises based on that narrative, studying how they enacted and played out the scenario narrative and its multiple events. This fieldwork was conducted mainly during the years 2015 and 2017 (as part of a research project funded by the Israel Science Foundation), when I closely followed the planning for three annual events (2015, 2016, and

2017), the writing of their scenario narratives, and the execution of the exercises themselves. Several research assistants provided valuable support that enabled me to simultaneously collect data from numerous sites during the time of the exercises.

Following this long-term project on national future scenarios, I began to inquire more broadly into the technology of scenario thinking and planning in two ways (as part of a second research project funded by the Israel Science Foundation for 2019–2022). First, I inquired into the expertise of scenario planners and the origins of this mode of thought and practice through an examination of the work of Herman Kahn and Pierre Wack and the scholarly literature about them. Second, moving from the national use of scenarios to their use at the international and global levels, I looked at how scenario planning is used within two leading international organizations: the World Health Organization (WHO) and the World Energy Council. Here, I was specifically interested in the ways in which these organizations chose and assembled their scenario narratives and the process though which these scenarios were delivered and experienced in their respective fields.

Accordingly, to further develop the conceptual framework of scenarios and uncertainty, I have examined the uses of scenarios in thinking about and preparing for unknown futures in the fields of energy and health. In my discussion of these cases, I show how future uncertainties are created and approached with the scenario technology. As part of my study of these questions, which began in 2019, I have thus far collected data from a range of different venues. The data on World Energy Council scenarios were collected from multiple official documents and the World Energy Council website, participation in the organization's scenario-planning workshops, and interviews. The data on WHO scenarios were collected mainly from official documents, the WHO website, and interviews. Planned participation in exercises had to be postponed as a consequence of the ongoing COVID-19 pandemic.

At the WHO, I explore and analyze the use of scenarios in exercises conducted within the organization. For scenario-planning activities, experts are appointed by the WHO's headquarters to construct multiple scenario narratives and events. These narratives and events can then be selected, reframed (to make them relevant to specific needs), and implemented and practiced at the regional (the regional offices for each of the six WHO regions), country (WHO country offices or countries themselves), or headquarters level, depending on the exercise's organizers. I examine how one regional office prepares for global emergencies using scenarios, looking at the kinds of scenarios that are chosen by the staff in the regional office and how narratives are selected and put into practice. At the

WHO, multiple scenarios (sometimes dozens) are written every year, relating to present, recent-past, and near-future problems (emergencies) and focusing on specific, known health issues (usually, but not limited to, diseases). Narratives also often target specific states or regions though the issues they address may have global implications. Interestingly, the scenario approach has been in use at the WHO for a relatively short period of time, and activities arranged by the organization tend to be what I define as simulation exercises, which are designed to test and rehearse the abilities of participants to respond promptly in predetermined ways to the types of potential problems contained in the exercise.

In contrast to its short history at the WHO, the scenario has long been used at the World Energy Council (and in the field of energy more widely) as a tool for assessing energy futures and suggesting ways of achieving long-term energy stability for the greatest number of people at the lowest environmental cost. At the World Energy Council, an internal team of scenario-planning experts is in charge of the triennial process of creation, revision, and update of narratives (only three scenarios at any one time) and their implementation in the organization, and the narrative-building process is initiated and guided exclusively by the team.[7] Each triennial cycle involves global, regional, and sectoral work. In 2016, for example, the World Energy Council published *World Energy Scenarios 2016: The Grand Transition* (World Energy Council 2016), which presented three scenario formats: "Unfinished Symphony," "Modern Jazz," and "Hard Rock." These global scenarios were designed to explore plausible futures for a world of lower population growth, radical new technologies, increasing environmental challenges, and shifts in economic and geopolitical power. More broadly, they provided the backbone, context, and framing for the scenario work during the triennial cycle. The World Energy Council's scenarios are built and updated annually, drawing on input from most of the institutions' members, and quantified using a global multiregional energy system model. The scenario process takes place in workshops conducted a few times a year with the participation of representatives from various states and organizations.

Extending the analysis from the local and national to the regional, international, and global levels makes available a broader spectrum of empirical material for understanding the technology of scenarios, including the processes by which narratives are constructed and other relevant practices. Taken together, then, the cases examined here provide a useful lens through which to view contemporary efforts to engage in an overall journey of discovering the future along with the modality of governing involved in these endeavors to face future uncertainties. Collectively, they enable us to understand in depth how scenarios express the new governing modality that I term *uncertainty by design*.

Uncertainty: Objects, Concepts, and Governmentality

Objects

Before discussing the sociocultural scholarship on uncertainty, let me briefly summarize the study of uncertainty within the social sciences more broadly. The identification of the problem of uncertainty in the social sciences is not a new development, nor is the conceptualization of risk and uncertainty as two distinct problems, according to which it is suggested that the former can be calculated and to a certain level predicted whereas the latter cannot (Bernstein 1998; Knight 1921). However, unknown futures—uncertainties—are recognized, conceptualized, and studied within most of the social sciences literature in purely ontological terms or as an ontological situation or problem—that is, as an event that exceeds our quantitative methodological capabilities (i.e., the science of statistics), as an extreme event that is impossible to predict, or as a complex problem in which the making of judgments and decisions is difficult.[8]

While statisticians have dedicated themselves to overcoming the problem of uncertainty by transforming the unknown into multiple possibilities that could each be known and controlled (see Lindley 2000; Tabak 2014), other scholars whose work originates in the same starting point of seeing uncertainty as a problem have proposed that academic research should be conducted to help identify possible biases in situations characterized by uncertainty or the impossibility of making accurate calculations about the future to enable improved responses to such circumstances.[9] This second set of approaches accordingly acknowledges both the problem of uncertainty and the absence of proper scientific methods for dealing with it, as well as the inadequacy of treating such uncertainty simply as "risk." As Kahneman and Tversky (1982, 143) put it, "philosophy, statistics, and decision theory commonly treat all forms of uncertainty in terms of a single dimension of probability or degree of belief." In place of such an approach, however, they suggest distinguishing internal from external attributions of uncertainty and sketch modes of judgment that people may adopt in assessing uncertainty—that is, in situations in which subjectively assessed probabilities are enacted. Put differently, they provide a framework for studying people's biases when they are acting under conditions of uncertainty.

Moreover, while some scientific approaches have taken a negative view of uncertainty, seeing it as a problem to be overcome (e.g., statistics or decision making), others have perceived uncertainty as an opportunity and have produced more optimistic accounts of the subject, noting that while uncertainty may create new spaces of risk, it may also offer considerable potential for creating advantages and profit (e.g., in economic contexts). However, going beyond these two different perspectives—that is, viewing uncertainty either as a positive or as a negative on-tological question—I begin by inquiring into the *epistemological* explanations of events in the world as either risk or uncertainty. That is, I ask how, in the first place, the future is observed through different conceptions (of risk or uncertainty) and how the application of the notions of either risk or uncertainty determines various distinct societal actions. Thus, rather than looking for new methods to overcome uncertainty or identifying the biases of people acting in situations of uncertainty (to eventually rectify them), I relocate uncertainty on a conceptual scale, viewing it not merely ontologically but also as a social concept.[10]

Concepts

Epistemologically, Niklas Luhmann (1993) suggests using the term *risk* to refer not to an object in a first-order observation (i.e., an ontological threat) but to a concept in a second-order observation.[11] From such a perspective, risk is a con-ceptual part of the social system and inherent in its decisions. Accordingly, Luh-mann (1993, 21) draws a distinction between risk and danger:

> To do justice to both levels of observation, we will give the concept of risk another form with the help of the distinction of risk and danger. The distinction presupposed . . . the uncertainty [that] exists in relation to future loss. There are then two possibilities. The potential loss is either regarded as a consequence of the decision, that is to say, it is attributed to the decision. We then speak of risk—to be more exact, of the risk of decision. Or the possible loss is considered to have been caused exter-nally, that is to say, it is attributed to the environment. In this case we speak of danger.

In this approach, the question is not about the quality of new dangers in the world (more or less severe, calculable or not), but about how the future is conceptual-ized in the present and contingent on it and how each decision concerning the future—or abstention from such a decision—determines risk. In the analysis pre-sented in this book (and elsewhere in my work), I draw on Luhmann's distinc-tion between risk and danger and propose an additional distinction with regard to the concept of uncertainty: I suggest taking uncertainty as a concept that reflects a way of observing and acting on the future, rather than seeing it as an object of the future. My interest, then, is in examining how uncertainty is inher-ent in the system's perception of the future and how the system's conceptualiza-tion of the future engenders uncertainty and operates in relation to it. Moreover, in developing the concept of uncertainty, I suggest distinguishing between two types of uncertainty: possible uncertainty and potential uncertainty.

Possible uncertainty, I argue, is comparable to the concept of risk. Any decision in the present regarding the future espouses one possibility as opposed to another. Action associated with this type of uncertainty is therefore linked to the knowledge that can be brought to bear on it, which is brought to explain, assess, and sometimes predict the future, within an ethos of controlling and reducing uncertainty by providing more knowledge regarding offered possibilities. Potential uncertainty, by contrast, derives from the variety of actualities that can emerge from the virtual event rather than from lack of knowledge about the content of any specific possibility. Gilles Deleuze's ([1969] 1990, 56–57) discussion of the relation between a question and its answers traces an analogous arrangement:

> The question is developed in problems, and problems are enveloped in a fundamental question. And just as solutions do not suppress problems but on the contrary discover in them the subsisting conditions without which they would have no sense, answers do not at all suppress, nor do they saturate, the question, which persists in all of the answers. There is therefore an aspect in which problems remain without a solution, and the question without an answer. . . . This problem does not at all express a subjective uncertainty, but, on the contrary, it expresses the objective equilibrium of a mind situated in front of the horizon of what happens or appears.

Potential uncertainty is like a question no answer can suppress or saturate. In this sense, potential uncertainty is not equivalent to the missing content of the unknown future but is linked to the intermediate space between what has occurred and what is about to occur—that is, to a particular form and temporality. In other words, potential uncertainty inheres in the relationship between the virtual and the actual. The virtual, Deleuze explains, "is opposed not to the real but to the actual. The virtual is fully real in so far as it is virtual." Whereas "the possible is opposed to the real," the virtual is opposed to the actual. That is, "the possible is open to 'realization,' it is understood as an image of the real, while the real is supposed to resemble the possible." However, "for a potential or virtual object, to be actualized is to create divergent lines which correspond to—without resembling—a virtual multiplicity." Therefore, "the nature of the virtual is such that, for it to be actualized is to be differentiated" (Deleuze [1968] 1994, 208, 211–212).[12]

Whereas possible uncertainty derives from lack of knowledge regarding the realization or nonrealization of a particular possibility, potential uncertainty derives from a state of virtuality in which various events can emerge simultaneously. Thus, whereas the former is connected to the lack of content (information or knowledge), the latter derives from a particular form that creates or designs this uncertainty (positioned between the virtual and the actual, and in both places at the same time—"in between," as Deleuze and Guattari would put it).[13] Possible uncertainty, then, is dependent on past knowledge, calculation, and evaluation (the chances of a particular risk being realized). It is comparable to risk, and various risk technologies are therefore available for acting upon it. Potential uncertainty, by contrast, does not derive from the question of whether one future possibility or another will be realized (as in the case of possible uncertainty) but from a virtual domain with the capacity to generate a broad variety of actualizations.

Governmentality

Moving from the phenomenon of uncertainty and its conceptualization, I also examine uncertainty as a possible rationality of governance. I suggest that we approach uncertainty neither just as something to be contrasted with risk nor merely as the outlier of risk-based rationalities—that is, as an irrational or extreme event. Rather, uncertainty may itself represent a new rationality, alongside that of risk, on the basis of which various governing technologies have emerged and been acted upon.

Science and technology studies criticize the (modern) techno-scientific approach as reflecting the idea that more knowledge about the future (risk) allows for prevention or control of its uncertainties. From such a perspective, Brian Wynne (2002, 468) identifies the need for "the cultural reification of risk in late-modern society." As Wynne (2002, 459) argues, "risk discourses have been fundamentally shaped by an assumption that any uncertainties which risk assessments might show will be resolvable by more science. The basic discourse of modern science and technology policy [assumes] that even if predictive control is not yet fully in our grasp, it soon will be."

However, in the age of "post-normal science" (Funtowicz and Ravetz 1991, 1993; Ravetz 1999), these scholars add, the world has become more uncertain, and risk assessment is not a sufficient solution for such uncertainty. For this reason, new techniques should be developed to improve scientific practice in the age of uncertainty.[14]

I see these approaches as evidence of a shift in perspective vis-à-vis the future: rather than being perceived as a space of knowledge-dependent possibilities that are manageable by means of science and technology, it comes to be seen as a space of increased uncertainty that, paradoxically, derives from additional knowledge and technological developments. However, as the governmentality approach set out by Michel Foucault ([2004] 2007) suggests, it is not sufficient for us, as scholars, to limit ourselves to a general narrative of transition from one grand social perception based on risk to another (i.e., one of uncertainty).[15] Contrary to what the "post-normal science" approach suggests, we cannot remain solely within the epistemological domain. Rather, we must provide an analysis of how forms or perceptions of risk (and uncertainty) produce particular modalities of governmental order in our societies and how these modalities are expressed in particular social technologies (see also Dean 1999; O'Malley 2004).

The focus on uncertainty shifts the discussion on modernity and governance from the risk society approach to the idea of potentialities both in security and in profit. If, as Giddens (1990) and Beck (1992, 2009) claim, modernization has constituted a state of perpetual uncertainty of a particular kind (what I term *potential uncertainty*), then it is impossible to understand practice, subjectivity, and governmental technologies without tracing the evolution of concepts of risk and uncertainty. However, instead of attempting to define a new era of uncertainty, I emphasize the importance of tracing the complexity of solutions that emerge in response to the broad problem of potential uncertainty and of examining how potential uncertainty directs a distinct form of governing.

In the governmentality approach, then, the term *risk* does not represent dangerous events or situations (or social perceptions of such events) but rather refers to a type of technology of governing and control. Thus, many studies discuss risk as a central thematic of the biopolitical security apparatus, grounding an approach to governing the population through the calculation and assessment of its conformity to or deviation from established welfarist norms. François Ewald (1991, 199–200) explains the rise of insurance in the nineteenth century in terms of such a technology: "By objectivizing certain events as risk, insurance can invert their meanings: it can make what was previously an obstacle into a possibility." Insurance thus uses risk in effecting a distinctive mode of governing, converting events into possible accidents that can be assessed and managed. Risk, then, "builds on the premise that [threats] can be classified, quantified and to some extent predicted" (Aradau, Lobo-Guerrero, and van Munster 2008, 147).[16]

Insurance is thus a technology of risk, but not because of the real danger it deals with. Rather, it uses risk as a technology of governing, as a way to deal with uncertainty, by converting it into possible accidents (Dean 1999; Lupton 1999). This rationality of governing, which turns something or someone into risk to make them governable, also appears in other areas of research, such as studies on populations at risk in Bogotá (Zeiderman 2016), old age (Kaufman 1994), psychiatry (Rose 1996), pregnancy (Lupton 1999; Rapp 1995), HIV/AIDS (Lupton 1994), and crime prevention and drug use (O'Malley 1992, 2004). In short, in risk technologies, future uncertainties are turned into a variety of possible risks, and in this way those uncertainties are managed and controlled.[17]

In developing such an analysis, Mitchell Dean (1999) identifies three main types of risk technology: the insurance type (risk spreading), the clinical type (reducing factors that might cause individual harm), and the epidemiological type (identifying and reducing societal risks). Pat O'Malley (2004, 7) similarly claims that the concern of governmentality studies is "primarily to understand risk as a complex category made up of many ways of governing problems, rather than as a unitary or monolithic technology." Following Dean and O'Malley, I suggest that one should also approach uncertainty as a complex category, a rationality of governing that is expressed in various contemporary societal technologies, including that of scenarios.

A number of scholars have discussed the notion of anticipatory governance (Adams, Murphy, and Clarke 2009; Anderson 2010a, 2010b; Osborne 1993), understood as a new form of governing that deals with risks and/or uncertainties that cannot be calculated or assessed (sometimes termed *nonrisks*). Some identify this form of governing as *preparedness* (Cooper 2006; Diprose et al. 2008; Lakoff 2008; Lakoff and Collier 2008; Samimian-Darash 2009, 2011a, 2011b; Stephenson and Jamieson 2009). The literature on preparedness argues that such an approach is directed toward almost unpreventable future disaster that can be managed only once it has happened. Interventions aim to reduce the damage from such developments rather than to prevent particular threats. Other studies use different terms to explore shifts in the form taken by attempts to govern the future (threat). Melinda Cooper (2006) has used the term *preemption* to conceptualize a mode of governing catastrophic future risk that denies the idea of prediction or representation. Rosalyn Diprose and colleagues term the new paradigm *prudence*, in which "the assumption [is] that the risks and threats are *incalculable, unpredictable but always imminent*" (Diprose et al. 2008, 269, emphasis in original). For their part, Claudia Aradau and Rens van Munster (2007, 102) have argued that a "precautionary dispositive" emerges "where the scientific technologies for 'representing' the world find themselves surpassed by reality itself."

Louise Amoore (2013) refers to the shift toward anticipatory governance in terms of "the politics of possibility" and identifies a moving complex assemblage that is emerging in the governing of uncertain futures through what she terms the "possibilistic mode." In a world of possibility, she argues, the economy becomes a means of securing future uncertainties, and economics seeks out security as a source of profit from uncertain futures. Thus, Amoore sees a correlation between the economic reason that approaches uncertainty as a potential source of profit and the security state or sovereignty power, hence the creation of a new mode of governing through the rationality of possibility (rather than risk and probability). This mode of governance makes future uncertainties possible (actual and real) to govern them. In consequence, various kinds of risk technologies emerge, such as private consultation, risk management, and software and biometrics engineering. Moreover, these risk technologies are marked by a new mode—that of possibility rather than strict probability. They do not govern by the deductive proving or disproving of scientific and statistical data but by the inductive incorporation of suspicion, imagination, and preemption. The idea here is that uncertain futures—however probabilistically unlikely they are—may be mapped and acted on as possibilities.

Although Amoore and the other authors mentioned above present new governmental technologies that obey rationalities other than that of risk, they still use the term *risk* in doing so. Amoore suggests extending the multiplicity of risk modality to include a possibilistic mode in addition to the probabilistic mode. In this book, however, I suggest distinguishing between the terms *risk* and *uncertainty* as well as between different rationalities and uncertainty-based technologies, as they express distinct modes of veridiction, jurisdiction, and subjectivation.[18]

In identifying this new governmental modality, I am inquiring into one particular uncertainty-based technology—the scenario—from an anthropological perspective, and I examine how it works on and in relation to the conceptualization of future uncertainties and efforts to prepare for them. Accordingly, I take scenario planning as my central object of research and observation and present findings from a long-term anthropological fieldwork research project on the micropractices of this technology.

The Future

Already in the 1980s, Haim Hazan and colleagues (1984) called for time to be studied in the way that one might study any social phenomenon. Similarly, Nancy D. Munn (1992) has explicitly posited time as a problem to be discussed separately from space and has urged anthropologists to recognize it as such (see also Gell 1992). For many years, however, anthropological study of time has focused more on the past and present than on the future (Munn 1992). Such oversight is seen as a result of the long-standing effect of the Cartesian dualism of tradition/modernity, which contributes to the discipline's historical focus on tradition and the more recent emphasis on the role artifacts play in the present (Bryant and Knight 2019).

Moreover, even as anthropologists have become more interested in the future, they have tended to investigate it from a perspective that prioritizes the present (Ringel 2016). Indeed, Sandra Wallman (2003, 2) has claimed that anthropologists are "less interested in prediction than in the causes and consequences of images of the future held in specific contexts of time or place" and in how such images may "affect what happens in the future" and even "constrain the present." In other words, futures were understood as contemporary conceptions and "imaginary presents" (Gell 1992).

Similarly, the philosopher Gilles Deleuze ([1968] 1994, 81–82) referred to the past as a form grasped only from the present: a "generalized past, therefore, does not exist in actuality, but is the *virtual* form of the past, accessible through varied practices of remembering," in the present (Hodges 2008, 411, emphasis in original). It can therefore be argued that the future can be perceived only as it is narrated or practiced in the present. Stine Krøijer (2010), who studied the protests of radical left-wing activists in Europe as a performance that affects time, has similarly claimed that the future appears in the present as a result of (present) performativity. In view of this presentist stand, anthropologists need to look at how future imaginaries are constructed and to reexamine the presumed linearity of past, present, and future to consider not only how the present affects the future but also how the future and different conceptions of it affect the present (Brigstocke 2016; Nielsen 2014; Peebles 2009; Zeitlyn 2015).

In this regard, Arjun Appadurai (2013, 285–286), who also argues that anthropology has turned a blind eye to the future, recommends observing the future as a "cultural horizon." Accordingly, it should be examined as a cultural fact variably constituted through three concepts: aspiration, anticipation, and imagination. Such an approach would enable an improved understanding of future-oriented ideas (e.g., prophecy, emergency, and crisis) across different cultures. Somewhat similarly, other scholars have inquired into specific cultural observations and images of the future, both as positive (Abram and Weszkalnys 2013; Cuzzocrea and Mandich 2016; Fischer 2014) and as negative (Jasanoff 2015; Lakoff 2015), making use of concepts such as hope (Miyazaki 2006; Rose 2007) and failure (Miyazaki and Riles 2005), as well as risk and uncertainty (O'Malley 2004; Samimian-Darash 2013; Samimian-Darash and Rabinow 2015), in their analyses.

More recently, however, scholars have discussed the centrality of the future in modern society and social perspectives and the proliferation of the topic in the social sciences in general and in anthropology specifically. According to Daniel Rosenberg and Susan Harding (2005), the Enlightenment's rejection of prophecy led to a form of demystification that transformed into alternative styles of engagement with the future. Following so-called modern rationales, new techniques for imagining and narrating the future proliferated, resulting in the current era being "boom times for the future" (Rosenberg and Harding 2005, 3).

In this regard, Jenny Andersson (2018) has studied the emergence of "the future" both as a social, political, and scientific problem and as a field of knowledge and expertise. She argues that the need to make the future knowable and governable has created an assemblage of various experts, scientists, and politics that needs to be unpacked. Andersson thus similarly takes the future as a social construct and asks how the future has become a social or political problem, to whom, when, and why. Going beyond this, however, I argue that the way in which the future is perceived and conceptualized (e.g., whether as risky or uncertain) also sets in motion specific cultural mechanisms geared toward handling the particular vision of the future adopted. Furthermore, through these technologies, the future is not only observed and understood but also created, realized, and designed. In addition, I highlight how by paying attention to the relationship between temporality and imaginary techniques seen in the context of scenarios we can not only see how futures are created and driven by the present but also redesign our presents.

Other work within the anthropology of the future suggests that such a direction merits further inquiry and analysis. Bryant and Knight (2019) provide a rich overview of the anthropological literature on the future, examining the roles and types of temporality in studying the future and how future orientations affect the present and present experiences. To them, it appears that the phenomenology of time has experienced something of a shift—an issue that has also been addressed in studies on the management (Bear 2014) and experience (Ringel 2016) of "modern time" (see also Kockelman and Bernstein 2012). Drawing on this perspective, Bryant and Knight (2019, 17) unpack time as "inherently teleological" in an "open-ended, indeterminate" way that is focused on "practices and orders of everyday life." They depart from previous anthropological works that approach time as linear by proposing "orientations" as a concept with which to approach the study of the future in anthropology. Actions in the present—and, indeed, the concept of the present—are dependent on the future and orientations toward it. Some of the orientations they discuss are anticipation, expectation, speculation, potentiality, hope, and destiny. Orientations such as aspiration and hope, for example, render the future malleable, while destiny and fate render it inevitable.[19]

The orientation of potentiality, they specifically suggest, encompasses that which is not actual: the elusive nearly-at-reach yet not quite present in the now, which is "being defined by being the other-wise-than-actual" (Bryant and Knight 2019, 180). That is, potentiality speaks for the not-yet-realized possible. In accordance with this logic, things may have potentials both to be and not to be, simultaneously. As it involves a state of being not-yet-actual, potentiality enables both surprise and a degree of expectation, and at the same time it may be completely invisible "precisely because unrealized" (Bryant and Knight 2019, 181).

In my proposed distinction between possible and potential uncertainty, I suggest these terms not only to argue about the indetermination of the future and its inherent openness but also to discuss various technologies of uncertainty.[20] That is, I offer different conceptions of future uncertainties for which distinct modes of interventions become possible. I furthermore address the specific technology of scenarios that emerges as a solution to the problem of governing future potential uncertainties. That is, in addition to defining the future as speculation, as potentiality, or in terms of any other type of orientation toward it, I am interested in the various governmental technologies applied to it and how they shape the very experience of the future by actually practicing, producing, and redesigning uncertainty in the present.

Scenarios

Scenarios are nowadays used within a multiplicity of fields, institutions, and organizations. From businesses to governments and nongovernmental organizations (NGOs), many have adopted the scenario technology and adapted it to their own specific needs (Ogilvy and Smith 2004). Scenarios thus play an important role in how the contemporary world is perceived and acted on, and their use has been studied in a wide range of fields and contexts.[21] In addition, scholars within sociology, anthropology, geography, and critical security studies have opened up a discussion on new modes of governing the future, and as part of this growing trend they have examined the use of scenarios as a specific technique within broader nonprobabilistic modes of governing that operate within the "politics of possibility" (Amoore 2013), using concepts such as precaution, preemption, premediation, resilience, preparedness, and anticipation.[22]

As James Faubion (2019) has pointed out, however, scenarios have mostly been examined in the specific context of security and emergency. Some of these studies have traced the migration of scenario techniques from US civil defense planning during the Cold War era into the fields of economics and energy (Cooper 2010), finance and law (Opitz and Tellmann 2015; Tellmann 2009), post-9/11 homeland security (Collier and Lakoff 2008), and public health (Lakoff 2008). In addition, many of them have looked at catastrophic, apocalyptic, "worst case," extreme, or disastrous scenarios.[23]

De Goede (2008b) and de Goede, Simon, and Hoijtink (2014) have shown how security practices of premediation deploy a limited imaginative capacity via scenarios to detect and disrupt disastrous virtual futures of terrorism, climate change, natural disasters, and pandemics before they occur. Tellmann (2009) has examined how scenarios of imagined future catastrophes are deployed in contemporary finance and insurance to produce new knowledge about risks. Opitz and Tellmann (2015) have described how within the fields of economy and law, scenarios inhibit an unpredictable, potentially catastrophic future. Similarly, Lentzos and Rose (2009) and Brassett and Vaughan-Williams (2015) have presented scenario planning as a future-oriented practice for preparedness within the discourse of resilience, which encompasses the notion of uncertainty as it imagines extreme, catastrophic events.

As the above indicates, when approaching scenarios as a form of governing, many discussions have made comparisons with risk-based technologies. Various scholars have drawn on Foucault's analysis of governmentality and his concept of biopolitics to argue that the intervention of power in scenarios does not depend on calculations of cost and expected damage, as in the case of risk. They have therefore looked at different ways of governing the future that go beyond probability and calculation: creative modes of governing that produce multiple plausible futures in the present.

In this literature, actuarial and statistical forms of governance that make reference to the construct of risk are replaced or at least challenged by a process of imaginative enactment (Collier 2008), by ways of attuning imagination to the unprecedented, unexpected, and uncertain, thereby exposing vulnerabilities and creating new knowledge about possible futures. These accounts, however, have suggested that scenarios make their appearance at the limits of calculation and risk-based rationality.[24]

In this book, I follow the idea that scenarios express a new mode of governing that is based on uncertainty and *goes beyond* the rationality of risk. Through analyzing how scenario planning emerged and shifted in response to the problems it encountered and sought to address, I also show how uncertainty—as a central element in scenarios—has gone from being an ontological problem for which the scenario technology (i.e., the use of imagination) was a solution to being an epistemological premise. I thus identify the role of uncertainty in scenario planning as being not just that of a mode of veridiction (which is how it is understood by Faubion) but also that of a mode of subjectivation. In other words, I see the scenario as expressing a new form of governing that has uncertainty as its mode of veridiction, jurisdiction, and subjectivation.

Other scholars have seen scenarios as a "future" technology (on futurism, see, for example, Andersson 2018)—a way to better know or anticipate the "unknown future" (see, for example, Aradau and van Munster 2007; Lakoff 2008; Lentzos and Rose 2009; Schoch-Spana 2004). In such a context, Andrew S. Mathews and Jessica Barnes (2016) argue that scenarios are part of prognostic politics. Scenarios enable predictions and action based on those predictions. They enable the imagining of plausible futures through planning for them and allow for new regimes of environmental forecasting. Susie O'Brien (2016, 329) calls scenarios "a strategy of speculation about possible futures." Similarly, Filippa Lentzos and

Nikolas Rose (2009) argue that although scenario exercises are not new in terms of their modality or action, they are novel in terms of their future temporality and their attempt to control uncertainty. I argue, however, that scenarios express neither a mechanism of prediction nor a risk-based rationality of calculation and assessment of future uncertainties. Rather, this uncertainty-based technology accepts the potentiality of the future, its uncertainty, and produces uncertainty in its modes of veridiction, jurisdiction, and subjectivation. Taken together, these various elements all reflect a form that I term "uncertainty by design."

Book Overview

In chapter 1, which discusses the problematization of scenario thinking in historical perspective, I trace the emergence of scenario planning and key shifts that have taken place in how the technology is used as well as how it came to represent a new form of thought and practice—that is, a new way of governing future uncertainty. I analyze two main approaches in the history of scenario thinking, those of Herman Kahn and Pierre Wack, and contend that scenarios first emerged in the work of the former as a unique means for thinking about the future through the use of the imagination. Subsequently, in the work of the latter, changes in the ways in which scenarios are used have brought about a shift in the technology, which has gone beyond simply being a way of confronting an external unknown future (event) to also promoting uncertainty as a mindset.[25]

In subsequent chapters, I present the three examples of scenario planning outlined earlier to show how the scenario technology approaches uncertainty in narrative and practice and how it shifts the experience of the subject in this process.

Chapter 2 examines how scenario narratives are constructed within Israel's NEMA. It explores the process and tools used to build a scenario narrative for national-level preparedness exercises, whereby the future is imagined and created to think both *of* and *through* it, as well as which stories are chosen for scenario narratives and to what end. My interest here is in how the real and the imaginable are intertwined in this attempt to make sense of the unknown future; how the imaginable scenario is kept reasonable—a challenging but balanced story that does not represent nightmarish or worst case eventualities; and how instead of predictions of "a future yet to come," the future is made present in this technology through the play with the *plausible* rather than the *possible*—that is to say, with stories, narratives, and imagination rather than calculations. This chapter thus explores the nature of the scenario's mode of veridiction, its form of knowledge making, and the manifestation of uncertainty as the logic of imagination in such a process. In this context, I suggest that scenarios represent a move from probabilistic approaches to "plausibilistic" ones in the governing of uncertainty—that is, a shift toward facing the unknown future with imaginable stories, a strong sense of realism, and acceptance of the uncertainty of the future as the basis for knowledge making about it.

In chapter 3, I explore how scenario narratives are put into practice by analyzing their social enactment and affect (i.e., their performativity) in the annual exercises organized by NEMA. In this context, I argue that not only do the content of the scenario narrative and the way in which it is built go beyond existing risk-based models but also the practice, the way in which the narrative is implemented, involves an approach to the unknown future that is different from that of risk, involving a mode of action that generates uncertainty. Although the scenario is usually built around a preselected and carefully constructed narrative, once it is implemented it is actualized as a multiplicity of subevents, or incidents, that are enacted by the various participants in ways that, in turn, often give rise to unexpected developments and processes of indetermination. In highlighting this phenomenon, I show how uncertainty is not simply the ontological problem that the scenario technology aims to address, nor is it merely the rationality that underlies the narrative-building process. Rather, uncertainty is both accepted in and generated through the actual practice of scenarios. Put differently, uncertainty is expressed in the scenario's form of action or power—its mode of jurisdiction. Thus, while knowledge making in the narrative-building phase of the scenario technology is based on the acceptance of uncertainty and the creation of stories yet to come, the practice, or form of power, of scenarios expresses uncertainty in the multiplicity of actualities that emerge from the scenario exercise during its implementation and in the generation of the unexpected that this entails.

In chapter 4, moving on from the discussions of veridiction and jurisdiction in chapters 2 and 3, I explore how scenario thinking and practicing also generate a distinct mode of subjectivation. Here, I also argue that the three modes of veridiction, jurisdiction, and subjectivation seen in scenario planning are mutually constituted and together express a form of governance based on a rationality of uncertainty. Drawing on observations from a long-term fieldwork project conducted at the World Energy Council, one of the world's leading energy organizations, I describe how scenarios are created by the World Energy Council's strategic planning department in a process that shapes not only new modalities of knowledge making and practicing in relation to the global energy sector but also new subjects, who are encouraged to accept and work with the uncertainty of the future through the creation of a new language, new frameworks, and new stories that narrate history from the future to help people address uncertainty by understanding first their own perspective and then other people's perspectives.

To date, the growing social sciences literature on scenarios has paid little attention to conceptual and analytical differences between diverse mechanisms of governing based on imagined future stories, which are generally lumped together under the umbrella term *scenario*. In chapter 5, however, through an analysis of preparedness exercises carried out by the WHO, I show how scenarios and simulations are distinct forms of governing based on different perceptions of uncertainty and the ability to control the unexpected. While scenarios are designed to address the emerging and the unexpected, the overall goal of simulations is to improve the use of predetermined practices and responses. That is to say, simulation-based exercises are designed to enable participants to practice responses from within a closed set of possibilities that have been defined and provided to participants in advance. Thus, while in previous chapters I have shown how the scenario technology addresses future uncertainty using a mode of jurisdiction (power) that promotes the unexpected and rejects the rationality of risk-based technology, what I show here is that simulations are based on different premises in terms of their form of action and practice: they draw on the past to create a closed set of possibilities and aim to establish an automated and preplanned response. There are thus considerable differences between scenarios and simulations in terms of the degree of openness involved in the two different types of exercise and the role of the uncertain and the unexpected in their practice.

Chapter 6 provides a broader view of scenarios and temporality. Drawing on the three cases presented in chapters 2 to 5, here I identify the different ways in which temporality is modeled in each case, and I examine the relationship between the specific scenario technology format used in each site and the corresponding role, perception, expression of, and concrete engagement with uncertainty. That is, rather than analyzing uncertainty as merely an object of the future or a problem "in the world" that needs to be solved or as a product of epistemology, a social perception or construct, here I have sought to emphasize how uncertainty is designed through the dynamic of each combination of scenario technique and temporality. What I attempt to show is how the temporality of a particular approach to scenario planning affects the contours of uncertainty observed in each case.

In the concluding chapter, I discuss the critical limitations of the scenario technology through an examination of the use of scenarios and simulations in the current COVID-19 pandemic. I show how the pandemic represents the actualization of a potential uncertainty event and how the shift in temporality entailed by this actualization affects the limitations of scenarios. Scenarios provide neither a best-prediction model of the future nor a preprepared guide for how to manage a particular situation correctly in the present. As the renowned scenario-planning expert Peter Schwartz put it, scenarios are "a context of thinking clearly about the impossibility" of the future—while, I would add, acknowledging its openness and potentiality.

In sum, this book is about a new mode of governing, one that marks a shift from risk-based technologies to uncertainty-based technologies in how we observe and prepare for the future (i.e., from probabilities and possibilities to imagination and potentiality). In addition, it inquires into how scenarios have become a central technology in efforts to prepare for future uncertainties. How has the scenario conceptualized and managed future uncertainties? What type of rationality is emerging as a result? I do not suggest that we should seek to idealize the shift that I identify in this work. My purpose is not to advocate on behalf the use of the scenario technology or to merely criticize the limitations of risk technologies, as the scenario has its own problematic externalities and critical limitations (Rabinow and Bennett 2012). Instead, my goal is to ask, what are the critical limitations of uncertainty-based technologies, particularly the scenario? And if risk-based technologies convert reality into tameable possibilities and thus provide the appearance of control, how does the proliferation of the unexpected in the scenario affect the reality of its participants and shape broader societal perceptions of the future in today's world? As a starting point for answering these questions, in what follows, I will set the historical context by looking at the emergence of the scenario-planning technology and the types of problems it was intended to address.

CHRONICITY
The Problematization of Scenario Thinking

In social scientific studies, scenarios have mostly been treated as a tool for imagining unknown futures (see, for example, Aradau and van Munster 2011; Krasmann 2015; Mathews and Barnes 2016) so that such futures might be practiced (through planning and preparations) in the present—a practice that should be distinguished from that of trying to *know* the future. An examination of the historical development of scenario thinking and planning, however, reveals that along with this new mode of veridiction wherein the future is enacted (Collier 2008) and which James Faubion (2019) has labeled "scenarism," scenarios also incorporate new modes of jurisdiction and subjectivation and represent a new way of governing future uncertainty.

In this chapter, I trace the emergence of scenario planning and key shifts in how this technology has been used since its inception. I analyze two main approaches to the scenario technology—those of Herman Kahn and Pierre Wack—and contend that scenarios first emerged in the work of the former as a unique means for thinking about the future through the use of the imagination. Subsequently, in the work of the latter, changes in the ways in which scenarios are used have meant that the technology has gone beyond simply confronting an external unknown future (events) to also promote uncertainty as a mindset, a way of thinking, experiencing, and practicing.

In much of his work, Michel Foucault (1990a, 1990b; see also Rabinow and Rose 2003, xviii) sought to conduct a genealogical study of problematizations—that is, a historical investigation into how the present contingent state came about, "to show how that which is so easily taken as natural was composed into the natural-seeming thing that it is" (Koopman 2013, 129). In this chapter, I draw on Foucault's conception of problematization to describe the history of scenario thinking and planning and to trace its emergence and evolution over time. As Foucault notes, problematization does not represent an already existing object or a newly generated object constructed by discourse. Rather, in reproblematizing, we reexamine "the ensemble of discursive and non-discursive practices that make something enter into the play of true and false and constitute it as an object of thought" (cited in Rabinow and Rose 2003, xviii). Thinking in terms of problematization in relation to this specific historical context, then, I ask how scenarios (contingently) emerged as a solution to the problem of future uncertainty.

In the light of this problematization, I show how the way in which the future was thought about had come to be seen as a problem, and how imagination and scenario thinking were promoted as a preferred solution to that problem. I suggest that scenarios emerged as a solution at a moment in time when a change was occurring in how the future was thought about and approached, and I describe how under Herman Kahn's methodology of scenario thinking, concerns shifted from the problem of *knowing* the future to questions about *the ways in which we think about it.* Accepting the fact that one could not predict or accurately know uncertain futures, Kahn promoted the use of the imagination as a method for rendering unknown future events thinkable in the present. In Kahn's work, scenarios would thus be used as a speculative framework, a machine to generate hitherto "unthinkable," "unimaginable," or unknown future events to make it possible to prepare for such events in the present.

Moving on from Kahn's pioneering work, I then show how through the later work of scenario expert Pierre Wack, scenarios underwent a further shift in relation to uncertainty: from being a technology that accepts the uncertainty of the future as part of its reasoning to a technology that actually generates uncertainty in and through its practice. While Wack was certainly influenced by the work of Herman Kahn, in his own work scenarios went beyond accepting (an external future) uncertainty and working with it, to become a way of producing a new *perception*—a means of encouraging managers to use uncertainty as an approach to decision making within their organizations.[1] Put differently, in Wack's approach, scenarios triggered a perception-changing process whose purpose was to make uncertainty a philosophical standpoint for seeing and acting in the world.

While various differences between the approaches of Herman Kahn and Pierre Wack have been highlighted previously (see, for example, Cooper 2010; Faubion 2019; O'Brien 2016; Tellmann 2009), in this chapter I show how these differences represent two different modalities of uncertainty within scenario thinking. In the first approach, scenarios emerge as a solution to the ontological problem posed by future uncertainty (working in the face of "existing" unknown futures)—a solution that endeavors to use imagination (rather than knowledge based on past information) as a form of reasoning about the future. This modality has been discussed by Faubion (2019), who describes the emergence of "scenarism" as a new mode of veridiction within a parabiopolitical form of governing. For Faubion, with its "scenaristic" rationale and "sophiology," the parabiopolitical represents a departure from biopolitics in that it replaces the statistical problematics of the population with the scenaristic problematics of the case. Rather than seeing it just as a technique to be used in the context of emergency and preparedness, then, Faubion defined scenarism as a new mode of veridiction related to this new parabiopolitical form of governing.

In this work, I will refer to this first strand of scenario thinking (i.e., the logic described by Faubion) as *Mode I scenarios*. In the second approach, however, scenarios appear as a solution to an epistemological problem of certainty about the future, a way of challenging and changing fixed perceptions about the future, of re-mediating one's perception of the world and accepting its uncertainty. In this second modality, scenarios become a way of entering into an uncertain sensibility and a particular mode of experience and practice related to and centered on the rationality of uncertainty—in other words, the scenario also becomes a way of thinking, experiencing, and practicing. I refer to this approach as *Mode II scenarios*.

Scenarios before Kahn

While the emergence of scenario planning is usually traced to the work of Herman Kahn during the Cold War, it is important to note that some central elements of the scenario technology had been in use within various fields and professional orientations prior to this time. Perhaps the most important of these usages, both for Herman Kahn's work and in terms of the scenario's historical importance more generally, involve the use of scenario-related activities in philosophical-literary works, military strategy, and theater.

Philosophical-literary works are related to scenarios in the sense that they provide concrete articulations of a future situation. Bradfield et al. (2005, 797) suggest that scenarios have been applied since early recorded history as "a tool for indirectly exploring the future of society and its institutions." Specifically, the authors refer to the works of Plato, Thomas More, and George Orwell as examples where scenarios are used in "the form of treatises on utopias and dystopias." While the topic is not discussed in any depth by the authors, they do suggest that such works explore possible futures in a very vivid and concrete manner.

Indeed, Plato's *Republic* was not merely a fantastic story but an outline or a blueprint for an ideal future society envisioned by Plato in the context of a debate about the future of Greek society (Pappas 2004). On the more dystopian side, Orwell's *Nineteen Eighty-Four* was not simply a novel but also an anticipation of a possible future, as were other literary works in this genre by authors such as H. G. Wells and Aldous Huxley (Claeys 2010). Utopian literary works by Sir Thomas More and his successors were attempts at political and theological critique through "not-impossible" visions (cited in Shklar 1965, 370) of developments that had not occurred in the past nor were likely to occur in the future.

Ideas from military strategic thinking—and particularly from the works of nineteenth-century Prussian military strategists Carl von Clausewitz and Helmuth von Moltke (often expressed in the form of war-game simulations)— were also key for the development of Herman Kahn's scenario planning, as Bradfield et al. (2005) have noted. Indeed, Kahn's *On Thermonuclear War* both resonates with and draws on certain ideas from Clausewitz's *On War* (perhaps one of the most influential works for contemporary military strategic thinking).

Clausewitz's work embodies the notion that in the course of war problems arise in an unpredictable fashion, making futile any attempts at exact prediction or calculation of a war's outcomes (Beyerchen 1992). Clausewitz induces the general characteristics of war by analyzing specific historical events. For him, war is a complex phenomenon that involves considerable uncertainty as aspects of policy, politics, and society combine in multiple ways with various planning approaches and the reality of combat, as well as the psychology and decision making of individual soldiers (Paret 2007). In the more practice-oriented parts of *On War*, Clausewitz (1982) applies the principles he has discussed earlier in the work and uses various historical examples to consider hypothetical war scenarios.

Prussia's military chief of staff Helmuth von Moltke incorporated similar ideas about the unpredictability and complexity of war in his writings on war planning (Holborn 1986). Such an approach was reflected in his emphasis on the need for flexible and rapidly adaptive strategies to face the uncertainties of war as well as in his delegation of tactical decision making to lower-ranking officers. But, perhaps most interestingly, it was also such an approach that drove Moltke to introduce the study of military history in military training schools (Rothenberg 1986).[2]

As the term *scenario* itself implies, this technique also draws on communication strategies that were originally developed for cinematic plot building. Although the literature on scenario planning often regards this as a mere anecdote, the Hollywood-inspired term *scenario*, according to John Ratcliffe (2000), was initially suggested by Leo Rosten to physicists at the American policy think tank RAND Corporation when they were searching for a concept to describe different hypothetical descriptions of satellites' behavior. In his sociological study *Hollywood: The Movie Colony, the Movie Makers*, Rosten (1941, 313–314) distinguished the work of "movie writers," or "scenario writers" from the traditional work of literary writers. Plot building or scenario building meant creating simple stories that allowed people to follow the plot of a production when imagery was missing. These plots were created as "mental bridges." Later on, once sound was introduced to movies, Hollywood producers started bringing playwrights and authors into the movie industry. Their stories, however, were not suitable for the big screen: "The men new to Hollywood wrote plays, not screenplays" (Rosten 1941, 315). Action—the movement of the story—was what was most important, and it was plot builders or scenario builders who were responsible for this element in films.

In her book on Cold War nuclear civil defense, *Stages of Emergency: Cold War Nuclear Civil Defense*, Tracy Davis (2007) uses performance theory to investigate the use of realism in emergency exercise planners' future scenarios. According to Anne Ubersfeld, the stage space is "the point of conjunction of the symbolic and the imaginary, of the symbolism that everyone shares and the imaginary of each individual" (cited in Davis 2007, 71). As Davis explains, the stage space may or may not refer to a real or existing space that the audience recognizes. In either case, the audience always considers what is happening on stage as unreal or untrue. Nevertheless, the more the illusion on stage resembles a known reality, the less it is experienced as real because it produces more "alienation towards the spectacle," while "the apparatus of theater" increasingly "comes into the spectator's consciousness" (Davis 2007, 72). In theater, then, fiction and reality are combined to create a complex interaction between them in a way that affects the crowd.

To be sure, the elements discussed above are central for both the historical development and the contemporary use of scenarios. That said, in this book I discuss scenarios as a wholly formulated technique that is intellectually linked to the work of Herman Kahn and his successors.

Mode I Scenarios: From How to Know the Uncertain Future to How to Think about It

At the end of World War II and the start of the Cold War, science and technology were growing increasingly complex. Concurrently, ties between state bodies and scientists that had been facilitated during World War II were strengthened through their increasing cooperation. For both sides, the aim was to tackle changes in the shifting political environment (Bradfield et al. 2005). Accordingly, an unprecedented alliance developed between the military, the state, and science, with the United States substantially increasing its investment in scientific projects for military use in this period. This juncture of the security state and techno-scientific knowledge was soon institutionalized through new assemblages and contexts of cooperation.

In 1946, on the heels of World War II, the RAND Corporation was established, engaging primarily with the problem of future war (future conflicts and defense). RAND was created to provide scientific solutions to security problems faced by the US Air Force during the Cold War (Hounshell 1997), particularly the need to secure the future of the United States in the event of a possible nuclear war (Masco 2014). In accordance with this initial mandate, experts at RAND were mainly focused on tactical management and decision making in the security field as well as on providing technical solutions in research and development related to weapons technologies (Bradfield et al. 2005; Chermack et al. 2001). Nevertheless, soon enough, RAND experts also began working on broader strategic aspects of security policy and attempted to develop a science of warfare, with objective methods that might promise victory in a future war with the Soviet Union. A central component of these efforts was the development of a technology termed *systems analysis* (Hounshell 1997; Kahn and Mann 1957).

Systems analysis sought to use various calculative tools and techniques (e.g., elementary economics, linear and dynamic programming, system simulations) as well as social science research (e.g., studies of Soviet economics, warfare capabilities, and decision-making processes) to optimize the potential for victory in a future war. For example, as one of the first studies conducted with the new approach, a systems analysis of strategic bombing sought to calculate the most cost-effective way of causing damage to the enemy (Hounshell 1997). The purpose of developing and applying systems analysis was thus to create knowledge about future situations of war and the optimal forms of action in such uncertainty-ridden situations (Kahn and Mann 1957). However, despite systems analysis's promise, RAND experts had to confront the problem of *specification*, which pertained to the specific circumstances under which future war might actually occur (Hounshell 1997). The questions thus became the following: What exactly should be optimized? Under what circumstances would it be necessary to carry out particular actions that required optimization? And if the future situation might change, how might we know what the optimal action would be, given that we do not know in which future we are going to act?

As a result, RAND experts were now required to decide which future situations were more probable than others. In response to this challenge, various techniques emerged that aimed to create realistic future situations in which actions would be required—namely, war-game simulations and exercises (Ghamari-Tabrizi 2005). One key approach that was developed as a way of addressing the new challenge was the Delphi technique, which was based on the assumption that bringing together a group of experts from a given field would reduce the margin of error in assessments on matters within their field of expertise. In the Delphi technique, experts were asked questions in a number of rounds and could revise their answers after reviewing the answers given by other experts (Bradfield et al. 2005; Ringland 1998; Tolon 2011). The developers of the technique believed that a quantitative synthesis of expert opinions regarding the probability of future events would provide a better prediction of the future—or of which future would eventually occur (Gordon 1994; Tolon 2011).

Both the Delphi technique and systems analysis sought to produce knowledge in response to the problem of the unknown or uncertain future. Systems analysis created knowledge on optimal courses of action in future conflicts but was not designed to produce knowledge or predictions about specific future situations that might occur (i.e., to describe what the next war would look like). In response, the Delphi technique aimed to create such knowledge on the basis of the combined calculations of expert opinions (and forecasts) about the future. It was against this background that RAND physicist and strategist Herman Kahn would develop his scenario approach. Rejecting prediction and seeking to open up multiple possibilities, Kahn (2009a, 182) suggested that "to appraise the future is . . . difficult because important aspects of the future are not only unknown but unthought-of. . . . An individual's view of the future is necessarily conditioned by emotional and intellectual biases. In addition, the future is uncertain in a statistical or probabilistic sense. There are many possibilities, and while one can attempt to pick the 'winner' of the 'race,' unless this choice is overwhelmingly probable it is more prudent to describe the probability distribution over the potential winners."

Kahn thus reconceptualized the problem of the uncertain future. Rather than seeing the question as being a matter of ways of *knowing* the future—that is, how to get more information in the present in order to know what the future will be—he saw it as being a matter of how to *think about* the future. Accepting the uncertainty of the future was not just about the fact that the future is unknown but also about how existing concepts and patterns of thinking render it "unthought-of."

Imagination, Kahn therefore suggested, could be used as a form through which one might think about different uncertain futures, beyond statistics or probability, beyond knowledge production (or drawing on the known past). As he further elaborated, "The effort of imagination and intellect required to bring a range of potentially relevant factors into focus is not likely to be wasted. . . . In particular, possibilities that do not seem live options today may become worthy of serious consideration overnight as a result of new developments" (Kahn 2009b, 156).

Working for RAND during the 1950s, Kahn developed scenario planning as what he initially termed "Future-Now Thinking" (Ringland 1998, 12), creating scenarios of nuclear war not to try to predict what would happen in the future but to examine different ways in which decisions, conceptions, and convictions held in the present might play out and unfold in various future possibilities. Initially, in the light of the many uncertainties stemming from the new and unfamiliar prospects of nuclear war, Kahn's scenarios were aiming at identifying uncertainties and bringing them into the space of present action.

While Kahn's colleagues' approach was to make knowledge in relation to the unknown future through prediction or probability, Kahn accepted the ontological uncertainty of the future and the impossibility of predicting it or knowing it in advance; in Kahn's (1965) own words,

> Trying to look, say, twenty-five years ahead is like trying in 1900 to predict World War I and the aftermath. . . . These things cannot be done, obviously. Nevertheless we are trying to look twenty, thirty, forty years ahead. And it turns out, whatever it's worth, it gets to be relatively easy to write scenarios in which you can see escalations occurring. . . . This is not in the sense of prediction. It's simply noticing that a lot of problems which somehow are not of great seriousness over the next five or ten years can mature . . . in the next ten, twenty, thirty.

Instead of prediction, Kahn proposed imagination as a new form through which one might think about and prepare for unknown futures in the present. In this sense, Kahn transformed the definition of the problem of the unknown future, shifting it from the question of (knowing) the future present (i.e., the present that will take place in the future) to that of (creating) present futures (i.e., futures brought into the present, imagined and realized as possibilities) (Luhmann 1998). While the former approach is engaged with predictions and knowledge of the unknown, the latter deals with mindsets and new ways of thinking about the future. For Kahn, imagination thus became a specific form through which one might think about futures—in the present.

In this context, Kahn maintained that dominant conceptions in thinking about nuclear war were based on wishful thinking rather than reasonable expectations (Bradfield et al. 2005; Chermack et al. 2001). As he once noted during an interview, "If you have a nuclear war it's not the end of history. You'll experience it, you'll live through it, and you'll have to answer afterwards for what you did during the war" (Kahn 1978).

Kahn's work on possible or plausible scenarios for the eruption of a nuclear war and its consequences thus marked a significant turn in thinking about the problem of the uncertain future. From Kahn's point of view, many of the people in his working environment (experts and policy makers) wrongly addressed the uncertain future by focusing too much either on sociohistorical continuity or on change and novel technological developments as the basis for imagining the future (Kahn and Wiener 1967). Experts using the former approach, he claimed, are "unimaginative," as they do not take into account the possibility that new precedents might be set in the future. Those using the latter approach, however, while relying mostly on new possible developments, ignored many important elements that would probably continue from the past into the "new" futures, and thus placed too much importance on possible far-reaching changes in the future when seeking to determine future policy.

Kahn's scenarios provided what Chermack et al. (2001, 10) have described as "detailed analyses with imagination . . . [to] produce reports as though they might be written by people in the future." Rather than attempting to accurately predict the likelihood of nuclear war or how exactly such a war might unfold, he put forward (imaginary) scenarios and tested how the existing approaches to nuclear warfare would unfold in each one (what he termed "forward thinking"). By doing so, for example, he showed how a "balance of terror" between the United States and the Soviet Union might lead "to calculated and miscalculated war" (Kahn 2011, 229–230).

Kahn accordingly argued that nuclear war is not a zero-sum game (in which you either destroy your opponent or are totally destroyed) and that one might survive a nuclear war and even win it. On the basis of such arguments, and because it was expected of him given his professional responsibilities (first as head of civil defense planning at RAND and later as the founder of the Hudson Institute, a future-oriented think tank), he recommended that the government invest in research and development related to civil defense along with ways of protecting the civilian population (Hounshell 1997; see also Lakoff 2007). Additionally, since outcomes that were highly unlikely or extreme were nevertheless possible, Kahn thought it was crucial to think about and plan for these by means other than calculating probabilities—that is, he provided a new mode of governing beyond risk and probability:

> Sometimes the best-laid plans go awry for "statistical" reasons. That is, a proper judgment may be made on the basis of the probabilities as they are known, but the improbable occurs; either conditions are met that

are far worse than anyone could have anticipated, or some bizarre combination of accidents—each one of which was unlikely in itself but could have been handled—takes place, and "swamps" the system. . . . Good planning is designed to decrease not only the likelihood of bad luck but also the consequences if it occurs, since the "extremely improbable" is not the same as the impossible. (Kahn 2009c, 209–210)[3]

In sum, in Kahn's approach, scenario planning emerges as a way of responding to the problem of future uncertainty by reconceptualizing that problem as the question of *how* to think about the future. Kahn's response aimed at imagining multiple possible futures in the present rather than predicting and calculating specific ones. Scenarios were developed as a form of thinking and governing that expressed a new rationality for approaching the problem of future uncertainty— as an ontology that could not be overcome or dismissed but rather should be imagined and played out in the present.

Mode II Scenarios: Generating Perceptual Uncertainty as a Mode of Subjectivation

Since the 1960s, the use of scenario-planning techniques (sometimes integrated with other future technologies such as forecasting and contingency planning) has spread widely. This proliferation was catalyzed by Kahn and other RAND experts (often referred to as "futurists") who left RAND and established new organizations and research institutes engaging with various policy and societal issues (see Andersson 2018; Tolon 2011). Over a period of several decades, scenario planning moved into new fields beyond those of security and policy, including the world of business (and later that of energy). At the Shell Corporation, scenarios soon developed into what became famously known as the Shell scenario-planning approach, with Pierre Wack, one of Shell's planners, playing a prominent role in this process (Kleiner 2003).[4]

The particular way in which Wack used and implemented scenarios within Shell brought about a change in the scenario technology. Where initially the technology had been used to think about the problem of the (external) uncertain future, scenarios were now used to encourage the acceptance of uncertainty, not just as an "external force" or reality but also as a way of perceiving the dynamics of the world.

In 1965, as part of attempts to plan its business future, Shell began using a computer-driven system that promised financial forecasting based on a rational model: the Unified Planning Machinery. This quantitative system aimed to "discipline" the company's cash-flow planning for a six-year period but was eventually rejected because its forecasts were deemed too short term and its predictions too inaccurate (Wilkinson and Kupers 2013). Shortly after this, Ted Newland and Pierre Wack introduced scenarios to the corporation (Chermack et al. 2001). Together, these two individuals built the "futures operation" at Shell, adopting the scenario technique as part of their work (Wilkinson and Kupers 2013). In 1967, a scenario-based planning project titled "Year 2000" was executed, followed by the "Horizon Planning" exercise in 1969. During these studies, the Shell planning team began noticing that their scenarios consistently pointed toward "discontinuity in the oil industry" (Bradfield et al. 2005, 799)—that is, the scenarios were suggesting that for the first time since the end of World War II, the oil market might cease to expand. Shell's management was introduced to this scenario and was therefore prepared for such a possibility. As a result, when the increase in oil prices finally occurred in 1973, the corporation was able to quickly adjust to the new situation (Bradfield et al. 2005; Chermack et al. 2001).

The challenge for Shell's scenario planners was different from that faced by Herman Kahn. Like Kahn, they believed that the future could not be predicted, and so similarly they aimed not to try to *know* the future but rather to *imagine* possibilities and practice them in the present. However, soon enough, Shell's scenario planners encountered a new difficulty: how to get decision makers within the corporation to accept that the future was uncertain and, in addition, to act with an acceptance of uncertainty—that is, not just to acknowledge the uncertainty of the external or ontological situation (the unknown future) but also to embrace uncertainty as a mode of practice and to act in accordance with such a perspective. According to Wack (1985b, 140), decision makers are often certain of their judgment, viewing the future on the basis of their own previous experience: "The usual scenario analysis confronts them [top managers] with raw uncertainties on which they cannot exercise their judgement. Because they cannot use what they consider to be their best quality, they often say, 'why bother with all that scenario stuff? We'll go on as before.'"

Presented with scenarios, Shell's decision makers were unable to apply the types of thinking they were accustomed to using. Uncertainties, as it were, went beyond the scope of their experienced judgment-based thinking. Wack therefore sought to change the way they thought. To do this, he modified scenario planning to address the problem he had identified: "The interface of scenarios and decision makers is ignored or neglected. By interface, I mean the point at which the scenario really touches a chord in the manager's mind—the moment at which it has real meaning for him or her" (Wack 1985b, 139).

In other words, Wack was "trying to manipulate people into being open-minded" (Ted Newland, quoted in Wilkinson and Kupers 2013, 122) or "changing the image of reality in the heads of critical decision makers" (Wack 1985a, 84). To challenge managers' perceptions of the future (so that they could accept uncertainty as part of their reasoning), Wack used scenarios to bring about a change in managers' thinking. As part of his efforts, he also involved decision makers themselves in the construction and carefully planned processes of the scenarios (see Chermack and Coons 2015).

Thus, while Kahn had developed scenario planning as an alternative to experts' attempts to determine what the unknown future would be and had proposed scenarios as a modality of governing uncertainty that drew on imagination in the place of calculations, Wack, working in a business context, was more concerned with the ability of scenarios to change how people think and envision the future: "We now wanted to design scenarios so that managers would question their own model of reality and change it when necessary, so as to come up with strategic insights beyond their minds' previous reach" (Wack 1985a, 84). Wack's future scenarios sought to take specific images of futures or specific mental models (what he termed the manager's "micro model") and challenge them by exposing these mental models to the bigger picture, composed of multiple future images (the "macro model"), in a process that would hopefully lead managers not only to understand and accept the uncertainty of the future but also to generate uncertainty as a form of thinking, experiencing, and acting. As Wack (1985b, 140) explained, "What distinguishes Shell's decision scenarios from the first-generation analyses . . . is not primarily technical, it is a different philosophy, having to do with management perception and judgement. The technicalities of decision scenarios derive from that philosophy."

However, as the scenario expert and strategic planner Ian Wilson (2000, 24) has argued, Wack's scenarios had to confront the issue that "by acknowledging uncertainty, scenarios underscore the fact that we cannot know the future, and so we perceive them as challenges to our presumptions of 'knowing,' and thus of managerial competence. And because few, if any, corporate cultures reward incompetence, managers have a vested interest in not acknowledging their ignorance, and so in resisting the intrusion of scenario planning into traditional forms of executive decision making."

In responding to this challenge, scenarios accordingly underwent another conceptual shift. In Wack's work they went from being a solution to the problem of how to think about or address the (external) uncertain future (an ontological uncertainty), and thus how to use imagination to do so without ignoring or eliminating multiple possible futures, to being a technology used to purposely generate a space for uncertainty as a mode of thinking, experiencing, and practicing.

To illustrate this difference, one can also compare two quotes that show that Kahn and Wack were concerned with different types of actor: Kahn, with the expert planner (such as himself); Wack, with decision makers and managers. For both, scenarios were designed to achieve a certain change in mental capacity (via imagination). However, the main targets, the *subject* of this mental work, is fundamentally different in the two cases. For Kahn (2009a, 182), "Since he can neither plan for, nor think of, everything, the planner presumably should try to look at a relevant range of possibilities, remembering the importance of examining possibilities which seem relatively unlikely but which would have very desirable—or catastrophic—consequences if they occurred. . . . Again the analyst, being less responsible for immediate decisions than the government official, but more responsible for 'stretching the imagination,' should, on occasion, be more willing to consider seriously the unlikely and the bizarre, or spend more energy in re-examining and reinterpreting the old and familiar."

Given that according to Kahn the future is an uncertain terrain, Kahn's scenarios targeted the imagination of planners or analysts (who are "responsible" for this mental activity). In this sense, as Kahn (1976) once argued during a debate, "None of us ever said we were designing our future. . . . We know we are making our future, we're not designing it." For Wack (1985b, 150), however, "a manager's inner model never mirrors reality; it is always a construct. . . . In today's world, a management microcosm shaped by the past and sustained by the usual types of forecasts is inherently suspect and inadequate. Yet it is extremely difficult for managers to break out of their worldview while operating within it. . . . By presenting other ways of seeing the world, decision scenarios allow managers to break out of a one-eyed view. Scenarios give managers something very precious: the ability to reperceive reality."

Wack's scenarios targeted the inner thought-constructs of decision makers. Puzzled by the "out-of-proportion" effectiveness of some "mediocre" future-oriented studies, Wack (1986) realized that it was necessary to change the inner "mental model" or "microcosm" of the manager. He shifted the focus from the expert (or planner) to the decision maker, and in doing so turned scenario planning into a technology for stimulating a change in the way reality was perceived by others. Wack thus went beyond recognizing future possibilities and imagining how they might unfold in different scenarios to effectively instill uncertainty as a philosophical standpoint in people's mind.

By drawing attention to the distinct problems that Kahn and Wack were seeking to address and how they used scenarios differently in response, I have highlighted the importance of uncertainty within the scenario technology and how it morphs in different settings. Indeed, it is precisely through uncertainty, which should be seen as distinct from the more common and widely discussed concept of risk, that scenario planning attempts to govern the future—in Mode I scenarios, by thinking about uncertainties through the use of the imagination; and in Mode II scenarios, by generating uncertainty as a mode of thinking, practicing, and experiencing. In the next chapter, I will look at a process in which the uncertain future is addressed by imagining and creating scenario narratives, in the specific context of Israel's Turning Point scenarios.

NARRATIVE BUILDING
Imagining Plausible Futures

The following paragraphs represent the beginning of a national scenario narrative plan that was written for the Turning Point 15 national preparedness exercise, the largest home-front exercise conducted in recent years in the State of Israel:

> It started in January. Scattered barrages of missiles—"dribbling," in military terms—most of them coming from the Syrian border, have been launched over Israel. For six months, during which Israel has conducted a policy of restraint, it has insisted on not being dragged [into a war]. However, during the past week the attacks [on Israel] have escalated [close to] the Lebanese border, and include the placing of explosive charges on the road system and attempts to harm IDF [Israel Defense Forces] soldiers.
>
> At the same time, rocket fire has also begun in the south. In the first days, these were isolated barrages that caused little damage and did not cross the Sderot line. Eilat also became a target after terrorist elements in the Sinai fired several missiles at it without causing significant damage. The intelligence assessment was placed on the desk of the defense minister, and on Friday afternoon the minister convened an urgent meeting of the HEEC [Highest Emergency Economy Committee], an emergency committee consisting of the directors-general (CEOs) of all governmental ministries; representatives of the electricity, water, and gas

authorities; as well as the police, the Bank of Israel, and, of course, senior members of the Home Front Command. . . .

"We understand that the situation is escalating, and it is very possible that we are on the eve of war," declared the minister of defense. During the course of the discussion, existing directives were refreshed and new directives were formulated; at the same time, the minister authorized the recruitment of several thousand reservists, primarily from the Home Front Command and the Air Defense System. The latest "round" [of attacks] began on Sunday. The cabinet meeting that morning was already overshadowed by the threat of war. . . . The prime minister gave instructions that the utmost efforts must be made to take control of events, to bring about the opening of the main roads that had been blocked. At the same time, he ordered the IDF to maintain restraint along the Lebanese border. . . .

On Monday morning, hundreds of advanced long-range missiles were launched over Israel, and many managed to evade the "Iron Dome" [missile defense system] and landed throughout the country. Most of the landings occurred in the north of the country. Residential buildings in major city centers were destroyed. Bridges on major traffic arteries were damaged. In some places, huge fires broke out. The air force began intercepting unmanned aerial vehicles (UAVs) that had crossed the border and were acting against the rocket launchers, while several Hezbollah cells managed to cross the border fence, some of which entered [Israeli towns near the border] and took hostages.

The IDF defined the situation as a high-intensity event, and the defense minister, in coordination with the prime minister and the IDF chief of staff, issued thousands of Order 8s [for the recruitment of reserve soldiers]. On the same day, a decision was made to evacuate tens of thousands of residents from the north. Absorption of the evacuees, it was determined, would be carried out by the Ministry of Interior's emergency division in coordination with NEMA [the National Emergency Management Authority]. (Ringel-Hoffman 2015)

The design of the plan reflects Israel's specific securitization context, encompassing both present concerns and historical processes. During its seven decades of existence, Israel has been involved in seven wars alongside the ongoing conflict with its Palestinian neighbors. As a result, the line between the external front and the home front is often a thin one, marking the latter as a paramount issue for the country.[1]

In its early stages, the scenario technology, as introduced by Herman Kahn (2011), addressed the question of how to govern the future in a new fashion. The problem was no longer how to know the future and thus how to create knowledge about it through various means (e.g., the Delphi technique or systems analysis). Rather, the question was how to *imagine* the future—how to think about the unknown future and prepare for it even as it remained unknown. Creating scenarios was a new way to use imagination as part of a future governance technology.

Sheila Jasanoff (2004, 2015) has suggested that notions of the social imaginary as elaborated in political theory (Taylor 2004) should be applied to what she refers to as sociotechnical systems. While I would agree that sociotechnical imaginaries offer prime points of entry for thinking about the way in which the present is governed, much of the work that has taken up Jasanoff's concept within science and technology studies has failed to specify the precise modalities by way of which such imagined futures get constituted and made actual. In addition, drawing on Crapanzano's (2004, 1) critique that "today's anthropologists have been less concerned with imaginative processes than with the product of the imagination," I would argue that we need to move from the notion of the national image or (social) imaginary to imagination practices (Sneath et al. 2009) and take this perspective a step further by asking how exactly processes of imagination occur and what imagination practices produce and enable. In response to the same critique, some anthropologists have reverted to looking at and developing ideas about the capacity to imagine at the individual level (see, for example, Schauble 2016; Wardle 2015). In this regard, Robbins (2010) posits that imagination provides the engine of newness and innovation because imagination is a realm of personal mental freedom unlike both perception and reason, which keep thought from constructing just any kind of world it wants.

This chapter explores the process and tools used to build a scenario narrative through which the future is imagined and created to think both of and through it. Here, I show the principles of the process of building the scenario narrative as they are manifested in Israel's Turning Point scenarios. I examine how a scenario is created and through which techniques, as well as which stories are chosen and to what end. My interest here is in how the real and the imaginable are intertwined in this process of making sense of the unknown future; how the imaginable scenario remains "reasonable"—a serious but balanced story (that does not enter into uncanny or worst-case realms); and how in this technology instead of a prediction of "a future yet to come," the future is made present through the play with the *plausible* rather than the possible—that is, with stories, narratives, and imagination rather than calculations. In other words, this chapter explores the nature of the scenario's mode of veridiction, its form of knowledge making, and the manifestation of uncertainty as the logic of imagination in this process.

Turning Point Scenarios

Preparations for each Turning Point exercise begin with an organizational directive (Series A) that determines the scenario's central theme and framework (e.g., war, terror, or earthquake) and identifies the various units that will participate in the year's scenario. The exercise is designed by National Directorate, a private security company hired each year by NEMA to test the entire state apparatus as well as NEMA's own readiness. Institutional participation in Turning Point has grown over the years. All ministries now take part along with all national infrastructure authorities (energy, water, etc.). Other participants include the Knesset (Israel's parliament), the State Comptroller, and the Bank of Israel. While most municipalities are fully involved in the exercise, others participate only virtually (i.e., a member of the exercise administration stands in for a given municipality and responds to events on its behalf). All participating entities are represented in the group responsible for the administration of the exercise (hereafter, "the administration"), and their representatives are responsible for training their respective units in coordination with the administration and in accordance with the general scenario instructions.

Once Series A has been completed, the administration commences work on producing the full narrative of the national scenario (Series B). The scenario will be based on an integrated perception of possible threats to the State of Israel, taking into account political issues in the Middle East as well as wider international issues. As Alon Kedem, a senior exercise administrator, explained to me, Series B is "the outline, the opening scene for the assignment exercise [Series C]. . . . It is a fictional story with a concrete purpose." Once a draft is completed, this outline (Series B) is distributed to all participants, and on the basis of this narrative, they then begin their preparations for the exercise. These include management role-playing exercises, simulations, roundtables, and interorganizational training programs. The Series B documentation provides the participants only with the story outline and the main themes and threats that will lead the year's scenario narrative. This is subsequently transformed into a more concrete and detailed story, Series C, through a months-long process that yields hundreds of pages documenting multiple scenario incidents.

This more-detailed story includes the specific events that will take place during the week of the exercise, which are tailored to the training needs of individual participating units (in response to their requests) and must also meet national objectives. Series C could be composed of thousands of incidents that are devised as "fallout" or consequences of the main events of the broader national scenario. Once they are written, the administration uploads these incidents to a central computer system—the "event generator"—and assigns each incident to the specific units or participants that it will affect, in accordance with their areas of responsibility. In time, during the week of the exercise itself, this detailed scenario will be actualized; that is, the system will activate the incidents of the events that the administration has written and will notify each unit of specific "occurrences" in its area of responsibility. Such incidents might include missile strikes, hazardous materials spills, terrorist attacks on schools, population and casualty evacuation, and damage to national infrastructure (e.g., electricity shutdowns) or emergency economy entities.

As Alon Kedem explained: "Series C, this is no longer a story. It's a lot of discrete incidents and news, not something whole—many split events, a lot of 'noise,' many 'junk' incidents that are not true, so that the participant should work to extract the [true] picture; so that the participant will learn to bring out the main points. Hence, Series C is a transition from the story to the isolated incidents that give the participant all kinds of happenings. Behind each event, there is a generator. A generator can be a missile, an attack, a cyber incident, a ground raid, a communication problem, etc. Events occur within an existing context."

Writing the scenario narrative is thus a process that takes place over several months, in which NEMA and the administration design the framework of the year's story and subsequently translate it into a detailed, concrete, rich story of multiple incidents. A war (or other emergency event) narrative is created, based on events and incidents that have taken place in the past, as well as new imaginary threats and events that are devised for the sake of the exercise. The developed story is thus distinct from anything that has already taken place, and at the same time does not seek to predict what is yet to come. *Plausible scenarios*—a term commonly used in this context—attempt to make the future present, and thinkable, by means of imagination.

Writing the scenario narrative is not a fully authoritative process, solely decided upon by the national administration (e.g., Adey and Anderson 2012). Drawing on comments and requests made by participants during the writing of the scenario, the administration aims to build a story that will incorporate numerous specific events relevant to individual participating units. In addition, the process involves a range of different techniques, including storytelling, graphs and charts, and computing methods and models and is overall a dynamic process.

Though the process of building the scenario narrative should not be separated from its practice over the week of the exercise, for the sake of explanation I distinguish between the narrative building and the actual implementation of the exercise, between the mode of veridiction and the jurisdiction of the scenarios involved.

Creating the Real through an Imaginable Story

In the early evening of 3 February 2015, I attended a dinner meeting at the home of Turning Point administration member Michael Yair. I was the first to arrive and found Michael in the kitchen preparing dishes for his guests. He told me he had invited the team assembled to write the 2015 scenario to help with the initial design of the exercise. Ten people subsequently arrived, and we sat around the dinner table on the porch and ate and talked. Those present were close friends, most of whom had known each other for a long time, and the atmosphere during the meal was relaxed, the conversation peppered with personal jokes. An administrative group would continue to meet weekly but would grow over time. By the week of the exercise, it would include approximately 250 people. The idea was that the administration would include representatives from each of the practicing units. These representatives flavor the narrative of the national scenario with their specific concerns and local observations, thus helping to translate the abstract scenario into specific events and cases to consider and practice.

When we finished eating, we reentered the house and sat in the living room on chairs and sofas arranged in a circle facing the TV screen. There, the lighthearted banter that had accompanied the meal gave way to serious, focused deliberation. The discussants represented the exercise operation team: the police, internal security, NEMA, the medical community, government ministries and local authorities, population logistics, and national infrastructure. These participants would oversee the practicing units during the week of the exercise.

Michael and Alon Kedem gave a PowerPoint presentation, which was a first draft of the story outline of the main exercise narrative. The two men presented about thirty slides, which consisted almost exclusively of verbal descriptions. No graphs. No charts. Tables, when they appeared, were text filled. Though just an initial outline, the presentation was comprehensive and richly descriptive. The first slide addressed key security trends in present-day Israel. Subsequent slides focused on international concerns and the political and military situation vis-à-vis each of Israel's neighbors (Lebanon, Syria, Jordan, Egypt, and the Palestinian Authority). Some slides described the attitudes of the Israeli public and media reactions to current concerns.

I was surprised by how seriously those in the room treated the presentation. Their attention riveted on the TV screen, they were not just going through the motions: they were engaged in reality. During the meeting, they repeatedly told themselves that their task was to "design a reality," that they were "a serious administration and hence need to create a scenario narrative that looks real," and that they needed "to get inside the head of the reality" and ensure that the scenario narrative that they would write would be as realistic as possible. At one point during the meeting, Michael approached me and said, "It is all a question of how to create a reality. All this imagination is for the sake of creating a reality."

The meeting went on for more than three hours. At one point, Michael summarized the general consensus: "Events of the scenario are a natural extension of [the real] reality. The reality of the exercise and the real reality should not be very far apart."

Whereas some scholars have described the scenario as being "haunted" by the "uncanny imaginable" (Masco 2006) and have suggested that the real is governed by the imagined, in the current case the imagined is "haunted" by reality. Imagination is used to create a scenario that is as real as possible. In either case, the imagined narrative and the "real" reality should not be seen as mere fabulation on the one side and the truth on the other. Rather, they are two coexisting realities: both are constructed and construct that which is real. As Krasmann (2015, 188) argues, anticipatory knowledge practices always constitute a fictive reality that is distinct from the supposedly real reality but is nonetheless real. That is, imagination is not the opposite of knowledge but is a different form of knowledge making. Put differently, imagination is not irrational but a fictive reality that is real.

Describing the process through which the narrative is written and how efforts are made to construct it as a reality or truth, Alon Kedem commented,

> You go and build the puzzle of the scenario narrative. You create the picture of the situation at the international level—the situation in Europe, the USA, and Russia—then you come to the regional arena: which states are the radical ones, who the allies are. You give the participant the feeling that this is the true (real) situation. Even visually, it will look real. We use real articles from the press.
>
> You want to help the participant to have a nice and friendly experience, and to make a commitment to the exercise. If I bring something archaic, they will not accept it. The goal is to bring the participant to this field with the maximum possible goodwill and motivation. This is the picture of the reality that they will practice when they enter the exercise and will have to play it out. They will have to disconnect from the real truth in reality and enter the new truth of the exercise in which they will have to make decisions. This is indeed a "scenario reality," but

it is still possible to reach depth and identification with it—identification with the story.

The narrative of the scenario includes concrete and realistic events that are relevant to each of the practicing units. These stories are initially based on accurate information; they draw on information initially provided by Israeli intelligence agencies, and refer to present and past events in Israel, both in real emergencies and in previous exercises. However, each year's narrative also breaks from this known information to create something new, a new imagined story that remediates the reality and the scenario narrative. This new reality, this imagined yet-to-be-realized narrative, is an important characteristic of the scenario, since it contributes to the real experience of the participants in the exercise as well as to keeping them involved and surprised every year and thus encouraging them to act in the exercise as though it were a real present situation.

Moreover, within the reality of this imagined world, fiction is applied not just to the design of the events of the scenario; time is also constructed according to the needs of the exercise narrative. Time can be slowed down and speeded up; it can be skipped and linked or sometimes reversed or duplicated. For example, depending on the scenario, sometimes hours and days are passed over so that the next stage can be quickly introduced. At one of the meetings during the narrative-writing process for the 2017 Turning Point scenario, for example, one of the organizers turned to the participants in the room and said that he was considering inserting a "time leap" in the narrative:

> On Tuesday at noon, we will leap five days forward in time. This is to say that during the first two days of the exercise, we will focus on the first days after the earthquake; then, on the third day, there will be a five-day leap, to the seventh day [after the earthquake]. The administration is considering changing the picture while keeping it "reasonable" and in accordance with the needs of the exercise. We will take into account whatever the Geophysical Institute presents us with, but the exercise will not simply be identical with this outline; rather, the outline will be adjusted to the needs of the exercise and the administration—while keeping events within a possible, reasonable scenario.

This designing of fictional time takes into account the "real" information provided by the Geophysical Institute about the possible fallout from an earthquake in Israel. Yet the administration aims to go beyond this pregiven information and to change the time frame of the scenario on the basis of what they imagine could happen after seven days of the event. That is, the administration creates a narrative that is both knowledge based (drawing on the formal information prewritten by the Geophysical Institute) and imagination based (created by members of the administration as part of the scenario event they want to practice, which is not limited to what is "real" or already known). These two coexisting realities enable the success of the exercise. For a serious play to take place, the knowledge-based reality needs to be taken into consideration, but for the exercise to be meaningful and challenging every year, the fictional also needs to be invented.

Though imagination and anticipatory thinking emerge at the limit of calculations and other forms of future knowledge making (Adey and Anderson 2012; Anderson 2010a; Krasmann 2015; Tellmann 2009), it seems that once the imagined reality has been created, predetermined forms of knowledge continue to compete with it. These competing modes are expressed, for example, in the abovementioned 2017 Turning Point exercise that practiced an earthquake scenario, as we see in the following description of events from that exercise:

On 11 June 2017, the first day of the exercise, I arrived at the Geophysics Institute around 10 a.m., knowing that at about 11 a.m. the national administration would unexpectedly start the exercise by activating (the scenario of) an earthquake of magnitude 7.1 in the Beit-She'an area of the eastern Jordan Valley. The situation room was open. A big oval table had been placed in the center, and two screens were open on each side of the room. A small table with coffee, tea, and some snacks had also been set out. Slowly, people from the different rooms arrived at the "situation room." At a certain point, they "heard" that the event had taken place, and all of those gathered outside began to enter the room and to take on a more serious attitude.

The only information that had been given at this stage of the exercise was that an earthquake of magnitude 7.1 had taken place. The participants were now "in charge" of the entire unfolding of this event. They needed to decide what exactly this meant in terms of how severe it was, which towns would be affected by it and to what extent, what would happen to buildings and infrastructure, what sort of death tolls or casualties had taken place, and other consequences. Using the assessments and initial information (from the Geophysics Institute and NEMA) about the multiple incidents that occur once an earthquake takes place, the rest of the governmental institutes and municipalities would further develop their understanding about the occurrences in their area of responsibility.

In a way, all other institutions participating in the exercise were waiting to hear from the Geophysics Institute, its preliminary situation picture and assessments, so that they could start making decisions and interventions in their areas accordingly.

At a certain point, the head of NEMA arrived in the room to get some first impressions of what had happened, seeking guidance on how to instruct the national level (the government) and the various participating institutions—basically, to get an idea of how bad the situation was. When he arrived, one of the experts from the Geophysical Institute took out a booklet that contained maps and information, already printed out and collected into one document. He gave it to the NEMA head and told him, "You can use these pre-written plans, which will tell you what will be the fallout of the [current scenario] event and any aspect you are interested in." That is, rather than beginning by imagining the "new" event and creating the reality of the exercise scenarios (that the administration had activated), the representative for the Geophysics Institute took out prewritten material and sought to examine the event of the exercise through the lens of these premade "known reality" plans. The NEMA head, however, kept asking the various experts in the room about how the imaginable reality they were supposed to be creating was developing: "I don't want to know what your assessments about a possible earthquake were, but about what is happening in this 'reality' [now-activated scenario narrative] during the exercise. Have you started gathering information from different units in the field? Have you started creating maps of damage?"

It took about two hours for the institute to deliver its first situation report. Only then could the rest of the participants begin acting out their roles and designing the imagined reality of the exercise from their points of view. Only once they had received this initial information could they start inventing their own individual local stories about what was happening in their areas of responsibility. In this case, then, there was a clear tension between the past-based possibility of reality and the imaginable reality, between whether to use what was written in advance or to create something new.

Though the exercise is based on narrative making through the use of the imagination, other competing models still exist. Moreover, although creating a new fabulated world, the administration emphasizes the importance of creating an event that would appear real. Only then, they reasoned, would it be possible to achieve full collaboration from the participating units. At one administrative meeting during the narrative design phase, a discussion centered on whether the outline that was being constructed was realistic and would be convincing in the eyes of the participants, causing them to act as they would in a real emergency. One senior team member and NEMA official, Yoel Lapid, said that "many people will say that it's not realistic," which led a colleague to ask, "What events should we add to make it realistic?" The point was to shape a credible story to elicit "real" reactions from participants. At one point during the same meeting, the administration head looked at me and said, "You see? We are designing reality. Amazing, isn't it? We are designing reality. A harsh reality." This combination of both designing something new and keeping a sense of realism was evident throughout the entire exercise.

It is not an easy task to keep this double reality a fictional almost-real reality. The boundaries between the imaginary world and the real are constantly blurred and challenged. In addition, changing the time and space of the scenario narrative could affect the severity and intensity of the narrative and hence lead to situations in which the scenario narrative could get "out of control." Accordingly, a balance has to be kept not only between the imagined and the real but also between the extremely challenging but plausible scenario and the worst-case narrative.

Neither a Future Prediction nor a Past-Based Possibility

Before administration officials create the national scenario (and its multiple incidents), they review information regarding potential threats to the state. As part of this process, Israeli intelligence agencies amass data, map possibilities, assess probabilities, and recommend a future threat that should be considered by the government. This creates a map of an "attributed aggregated threat" (a probable future threat) that is distinguished from what will eventually become the "attributed scenario" event created for the current year's scenario. On one occasion, Yaron Weiss, a NEMA official, explained to me that "the attributed scenario is the basis for the work plan for the exercise. In other words, it is actually a decision. It is a decision of someone who says, I confirm . . . that in our action plan we will prepare for X missiles and not Y missiles. . . . The intelligence information [on which the attributed threat is assessed and gathered] . . . from the moment we decide that we are going to use such a scenario threat, actually has no meaning, because we have decided that this would be the scenario, so now we are preparing for the scenario, and the intelligence information and probabilities are no longer relevant."

During preparatory meetings, I repeatedly asked about the relations between the "real future threat" (which has yet to occur) and the "attributed threat" (practiced in the scenario). I was consistently told that the important issue is not the "real" future threat "out there" but the reality of the scenario. Although officials create a scenario on the basis of the attributed threat constructed through intelligence gathering, the scenario is not aimed at predicting the coming threat or the "real" future threat. A scenario narrative is chosen. That is, whereas the attributed threat as a calculation of the possibility of future events aims to predict or replace the unknown future, the scenario narrative draws on this information but goes beyond it and re-mediates it. Since the task is not to predict or to realize the future before it takes place, the administration chooses a particular story (or stories) to develop the work of preparing for and with uncertain futures without prediction, accepting the potential and the openness of the future without any form of reductionism. The following extracts from my field diary demonstrate this point:

> 1 June 2017. Preparatory meeting for an earthquake preparation exercise at NEMA. The administration staff discuss the information they have about earthquakes in Israel and which scenario to produce for the year's exercise.
>
> A representative from the Geophysical Institute explains how the Institute evaluates, calculates, and prepares for such events: "We have no model calibration. We have a model with a great deal of uncertainty. While this is not a guess, there is considerable uncertainty in these models. So our staff took the Turkish model and, because of its proximity and geological similarity, they tried to predict what would happen in Israel. . . ."
>
> A representative from the administration explains to the participants: "I told you beforehand: there will be no congruence between what the Institute says and what we are going to do [the Institute tries to predict the results of the future earthquake, and the scenario has no such intentions]. We have a tendency in the administration to get details of the scenario up to the level of the casualty. But we must not try to control every number, because in reality things will not happen that way. There are no exact facts [or numbers]—the reality is much more chaotic."

I once asked a member of the Turning Point administration to articulate the objective behind the choosing and narrating of the scenario event, and he answered that it was to "simulate reality as closely as possible." I pressed for clarification: "As close as possible to actual future reality?" He responded, "No, to the scenario reality." In other words, not only is the attributed scenario distinguished from the future attributed threat, it also does not attempt to predict the (yet to come) future but seeks rather to create a plausible future present, one the administration can choose according to its needs, to think of and practice that eventuality. While the attributed threat aims to present the "real" uncertain future, according to knowledge based on past events, the attributed scenario is separated from the future prediction of threat and is not merely driven by what has happened in the past.

Scenario narratives have been sometimes presented in the scholarly literature either as prophetic or as a way to better know or anticipate the "unknown future" (e.g., Aradau and van Munster 2007; Lakoff 2008; Lentzos and Rose 2009; Schoch-Spana 2004). However, as I see it, the temporality of scenarios is not merely "the future." Instead, it lies between the future (or the "as if") and the present (Anderson 2010a), between "future presents" and "present futures" (Luhmann 1998). In this unique positioning, the scenario narrative is neither a future prediction nor an already known (past) possibility.

As Krasmann (2015, 187) puts it, the gap between "the present and the future cannot truly be overcome by our faculty of foresight." In other words, imagination practices emerge at the limits of knowledge based on predictions or past-based information, owing to the time gap. Moreover, Collier (2008, 225) adds that, via scenarios, knowledge about the future is sometimes produced through the practice of "acting [it] out." Thus, the scenario is not simply a realization of the future but a way of creating new knowledge about the future by imagining it and "practicing" it.

Hence, instead of normative scenarios aimed at predicting or reaching a desirable planned future, experts within Turning Point explain that their scenarios are based on plausible futures and are viewed as reasonable because they are seen as worth practicing rather than because they are believed to represent the exact or true future to come. In this regard, Eyal Barak, an official responsible for writing and directing the exercise, explained to me that scenario narratives are selected, chosen, rather than predictions: "We are talking about a serious and reasonable scenario. This is what we design—not a worst case, but plausible and serious. [Moreover], I'm not saying this is what's going to happen, but that this is what we need to prepare for. This is not a prophecy. It's a decision, and we need to be ready for it."

Plausible Scenarios

At 2:00 p.m. on 2 June 2015, during Turning Point 15, ten people were gathered in the situation room of the Israeli Ministry of Transport. The walls were covered with lists of telephone numbers, incident schedules, and maps of key transportation routes. Two large-screen TVs hung from one wall, their data displays continuously updating. The situation-room staff sat around a desk monitoring laptops and telephones. As information came in from a variety of sources, the staffers relayed it to one another. While no one appeared apprehensive, the sense of pressure was palpable. Concentration was written across the faces of all the staffers as they struggled to keep up with the incoming reports.

> STAFF MEMBER 1: Rocket attack in Haifa port. Two containers hit. Another rocket hit the passenger terminal. A fire broke out, and people were injured.
> STAFF MEMBER 2: So, is the port closed?

STAFF MEMBER 1: There is damage, but the port is still functioning.

STAFF MEMBER 3 (looking at a TV screen): There's a bombing in the Azrieli Towers!

STAFF MEMBER 4: There's another report coming in: A direct hit on an ammonia tank!

REACTION IN THE ROOM: Thousands could die . . . that's many kilometers of damage . . . population evacuation.

STAFF MEMBER 1: There's a warning regarding a cyberattack.

MEDIA REPRESENTATIVE (playing a news flash): One hundred and twenty killed. Rocket hits. Many casualties. Number unknown. A group of terrorists invaded from the shore. Reserve forces mobilized. In many areas, the sirens are inaudible.

STAFF MEMBER 2: OK. There is a new event coming. . . .

REACTION IN THE ROOM: WHAT?! A new event? We aren't done dealing with the previous ones!

The scenario for Turning Point 15 was unprecedented: 24,000 incidents were written in advance, representing an all-front war. It presented the country with an emergency situation unlike anything it had experienced before. During the months of narrative writing, however, the administration officials insisted that their purpose was not to create a worst-case scenario: "We shouldn't overdo it with the number of events [requested by participating units]. Let's not overdo it. It's a blow, *but not an unreasonable one.*" Though acknowledging that the scenario involved an extraordinary event, authorities did not conceive of it as a worst case. A worst-case scenario, in their view, was fiction, impossible. A scenario, they contended, should be "serious but reasonable," situated between what is known and has already occurred and an imaginable eventuality that is challenging yet manageable. Yaron Weiss argued,

> [On the one hand,] it can't be something abstract, too high . . . that I can't reach in terms of preparedness. Otherwise, it's just a Hollywood script; it's not a basis for a work plan. On the other hand, it can't be built in such a way that I already know how to deal with it, because then I'm not reinforcing my ability. . . . The attributed scenario should be what is called the serious plausible level. It's serious enough so that I need to prepare for it . . . and be proactive, but it's plausible so that I have the option to reach it and deal with it. So the game is not where I am now, but it's also not in a place I can never reach.

In one of the meetings of the administration, during the writing of the scenario, the staffer who represented the point of view of the police explained that the severity of the narrative should be controlled so as not to "break" the participants:

> There are events that will bring us to a situation of a worst case that we want to avoid. You have to think about which events you can or cannot put into an exercise to create an escalation. . . . We should not break the police—otherwise the participants will not "play." It is forbidden to bring the police into a state of loss of control. Its peak will be on the second day of the exercise, and even then, it must not be broken. . . . We should not produce a national catastrophe, but escalate what exists [to a certain limit].

However, there were many moments during the exercise that participants experienced the event as going beyond their capacities. In another episode during Turing Point 2015, the tense atmosphere created by the scenario narrative could be clearly observed:

> 1 June 2015. Ministry of Economy. 3:00 p.m. The words "I don't know, I don't know, I don't know" are thrown into the air. There are just a few people in the room, five men and one woman. Everyone is working non-stop. The ministry consultant on emergencies tells Ronen Avraham from the Emergency Human Resources Authority that there had been no situation assessment that morning but that information is now flowing again.
>
> 4:00 p.m. The situation evaluation starts. Raol Esmaill, a senior representative from the Ministry of Economy, speaks first, saying that today is the first day of war. The Ministry of Defense still has not issued an emergency directive. We are at war, but it still has not officially been declared. There is uncertainty. There is no information about what is going on. NEMA has issued a population evacuation directive affecting approximately 30,000–40,000 people along the northern border. It is important that the district managers know where these people come from and where they are being evacuated to. The district reports begin. There is constant noise in the room.
>
> The Northern District reports that all daycare centres are closed. Those present also discuss the possibility of price gouging, and the Northern District representative asks that controls be strictly enforced. Raol Esmaill says angrily that it is impossible to activate all necessary procedures on Day 1. You cannot activate every emergency procedure without an apparent need to do so. Tzroya Haggai from the Tel Aviv District talks about rocket hits in three locations—the Central Bus Station, the Azrieli Towers, and another location.

4:25 p.m. Ronen Avraham speaks: "There's a mess in the system. There are many events. There's a mess online. They are confusing us" [i.e., the exercise administration is sending multiple events simultaneously]. He's upset. "I don't know where I'm at [regarding Emergency Human Resources]. There are no data regarding factory workers arriving at work. We have received information from NEMA but not from the districts. . . . This is a drill, but tomorrow this could be the reality."

The person in charge of emergency food reserves [at the Ministry of Economy] speaks. She is upset because she was asked whether the reserves should be opened. She says it is unnecessary, that this is just the first day. There is adequate stock. Everything is okay. She is not opening a thing. And then she [a Romanian immigrant] says, "Even during Ceaușescu, where there really was no reserve, I didn't see such hysteria like in this exercise."

In the scholarly literature, scenario narratives are often associated with worst-case and apocalyptic visons. Studies have examined catastrophic scenarios and apocalyptic threats, along with mechanisms of both governing them and governing through them (Anderson 2007; de Goede and Randalls 2009; Masco 2006; Tellmann 2009), representing scenarios as worst-case techniques (Clarke 1999). Monica Schoch-Spana, for example, argues that bioterrorism-response scenarios operate in a milieu of emergency and apocalyptic future perceptions, suggesting that scenario narratives are generally written as worst cases in which "everything that can go wrong does go wrong" (Schoch-Spana 2004, 12). Joseph Masco has described a similar approach in his studies of Cold War scenarios and the "nationalization of death" (Masco 2008) and biosecurity threats (Masco 2014). However, as several authorities explained to me, Turning Point scenarios are selected and designed to be serious and challenging but not beyond reach. Similarly, Ben Anderson and Peter Adey (2011) found that in the United Kingdom, specific scenarios are chosen over others not because they represent an apocalyptic future but because they are reasonable.

In the preparations for Turning Point 16, Michael Yair reflected on a worst-case scenario in these terms: "When I build a scenario for an earthquake, I need to build it so that the people who are training feel that they are able to deal with the scenario, and I won't always give them the most extreme exercise. I will [even] aim a little lower than the possible. Sometimes, even that could be too much." Eyal Barak told me that during the implementation of the exercise his job was to control the limits of the scenario so that it would not exceed the original scenario plan and become impossible. This control, he asserted, was essential for the success of the exercise, which he contrasted with Operation Dark Winter, which has been conducted in the United States in 2001 as a worst-case event:

> Six months ago, I prepared a huge Excel table that determined . . . the rocket trajectory, how many buildings would be damaged, how many people evicted, how many injured, how many people with special needs would result from this, and how many injuries—serious, light, etc. . . . Everyone worked according to this table. . . . Otherwise, suddenly anyone can add injuries and we have no control over it. I mean the control is, again—we wanted to set the number of injuries in specific regions so that the health system in some places can't deal with it, in other places it can. For each detail, we thought what . . . we want the trainees to gain from the exercise. Because if we collapse all of the systems, and hit all of the hospitals . . . this is what happened in Dark Winter, they just chose the extreme and that's it [everybody died].

The scenario addresses future uncertainty by creating plausible future stories and putting them into action in the present, not for the sake of prediction or making accurate prophecies but to act out new knowledge about the unknown future. Notwithstanding, keeping the scenario plausible is also dependent upon the provision of a story that is challenging yet not impossible to think of and prepare for. That is, these narratives are not worst-case scenarios; on the contrary, if they were too extreme, that would shift them from the real and reasonable to the unreal and impossible.

These narratives of plausible futures enable the creation of a road map not only from the present to the future (from the past into prefuture predictions and assessments) but also from the future to the present: the imagined future now governs the present (see, for example, the discussion of premediation in de Goede 2008a). They enable thinking about the unknown future in the present to improve present actions toward it. In other words, instead of governing the future through risk and other (temporal) figures or orientations based on the past, the issue is one of governing the present by or with future models and figures or logic that are based on the use of imagination. That is, approaching the problem of future uncertainty by creating imagined plausible future scenarios not only governs the future and reproblematizes the contemporary situation in the light of future orientation and concerns (or at least reflects that trend) but also regoverns our present and makes the present an artifact or an outcome of future observations and actions.

Uncertainty-based thinking toward the future has emerged at the limits of the governing of the future by reducing the latter into a present future (as risk technology does). To achieve this, scenarios, in particular, are based on plausible future stories, where the aim is not to attempt to predict the future—the unknown future—in advance or to reduce the future into these specific stories. Rather, the stories create an imaginable reality that couples with the "real" reality yet goes beyond it to create something new.

EXERCISING

Practicing the Unexpected

In chapter 2, I showed how scenario narratives are chosen, created, and developed and how they mark a move in the governing of uncertainty from probabilistic approaches to plausibilistic ones (see discussions of the "possibilistic approach" in Amoore 2013; Clarke 1999). In other words, scenarios represent a move toward facing the unknown future with imaginable stories, in which a sense of realism is predominant and where acceptance of the future's uncertainty is the basis for knowledge making about the future.

In this chapter, I would like to extend my argument about scenarios to illustrate how it is not just the actual content of the scenario narrative and the way in which it is produced that go beyond existing risk-based models; rather, the practice, the actual execution of the narrative, also involves a different approach to the unknown future along with a mode of action that in itself generates uncertainty. Although the scenario is usually constructed around a preselected and well-designed narrative, once it is implemented or put into practice it is actualized as a multiplicity of subevents, or incidents, that are enacted by the various participants in ways that, in turn, often give rise to unexpected developments and processes of indetermination.

My goal here, then, is to show how the imaginable yet realistic stories that make up the scenario narrative are put into action during an exercise and how the scenario technology manifests uncertainty not just in the premises of its narratives but also in its actual practice. In doing so, I will show how uncertainty is not simply the ontological problem that this technology aims to address. Nor is it merely the rationality that underlies the narrative-building process. Rather, uncertainty is accepted in and generated through the actual practice of scenarios. Put differently, uncertainty is expressed in the scenario's form of action or power, that is, its mode of jurisdiction. Thus, while knowledge making in the narrative-building process of the scenario technology is based on the acceptance of uncertainty and the creation of stories yet to come, the practice, the form of power, of scenarios expresses uncertainty in the multiplicity of actualities that emerge from the scenario exercise during its implementation and in the generation of the unexpected that this entails.

Viewing preparedness exercises as a technology that makes it possible to anticipate future unthinkable, incalculable events (Anderson 2010a; de Goede and Randalls 2009), scholars have paid considerable attention to the ways in which knowledge about the future is generated in them (Anderson and Adey 2011; Aradau and van Muster 2007; Collier 2008; Cooper 2010; Lakoff 2007; Schoch-Spana 2004). So far, however, such attempts have not provided a nuanced analysis of *how* the improbable and the unknown are approached, foreseen, imagined, or addressed (as a problem) in the exercising of such scenarios; in other words, not only how the scenario functions as knowledge making of the future (i.e., its mode of veridiction) but also how the unknown, unpredicted, and uncertain emerge through the performative practice of the scenario technology—that is, through the actual implementation of an exercise (i.e., its mode of jurisdiction).

Practicing the Imagined, Experiencing the Real

In Turning Point exercises, the exercise's scenario might start as much as two weeks after the occurrence of the hypothetical emergency event that forms the central theme of the scenario narrative (e.g., the outbreak of war or an earthquake). At this time, governmental offices begin to receive alerts about multiple concrete incidents that have taken place in their areas of responsibility as a result of the emergencies that have erupted. The government's emergency headquarters is opened, and all practicing units (bureaucratic organizations) are required to open their situation rooms and gradually integrate into the flow of the exercise. It is at this point that notifications of thousands of small incidents prepared by National Directorate begin to be transmitted to the various participating organizations, according to their specific locations and areas of responsibility. These notifications are communicated through various means, particularly via the "event generator" (a computer program) but also via e-mails, phone calls, and radio and television reports. Information about the various incidents that have occurred is gathered by each participating unit and then aggregated and analyzed in their individual

FIGURE 3.1.   Emergency situation room during Turning Point exercise.
Photo by the author.

situation rooms, which are generally characterized by an atmosphere of emergency and urgency (see figure 3.1). After the various local units or organizations have gathered and analyzed this information, they report on what they have discovered to their higher authorities, with this information eventually being passed on to the national situation room.

On entering a situation room in a government office during this time, then, one might hear a buzz of voices and see people hunched over computer terminals, huddling together for discussions or taking a few minutes at the side to take notes and reflect on how they might respond to the events that have occurred.

On 11 June 2017, during the Turning Point exercise carried out that year, which was based on an earthquake scenario, I visited the Ministry of Health's situation room. The room was huge, with a very long table that was mostly empty. Three groups of people were at work around it. Most were women, though there was one group of three men. Other people were constantly going in and out of the room. Along the walls were two large projectors and seven large TV screens. The screens displayed e-mails of incoming reports, maps, or the emergency exercise system. On a whiteboard was a report of an ammonia storage tank that had tipped over.

One of the attendees asked whether there was a problem in Jerusalem. Someone else in the room answered that everything was okay and that the hospitals were still functioning. Another message came in, forwarded through WhatsApp, announcing that the Haifa blood bank was out of order and that this information should be forwarded to Logistics. I left the main room and went into the other rooms located around the main situation room, each housing a different branch (Pharmaceuticals, Hospitalization, Logistics, Home Front Command, Psychiatric, Volunteers, Information, and others). In every office, the occupants looked very busy and were working in a very serious manner.

At 1:55 p.m., someone announced over a megaphone that everyone should come to the main table within the next five minutes. At 2:00 p.m., a situation evaluation meeting started. Once everyone had seated themselves around the table, the emergency preparedness chair of the Ministry of Health, speaking into a microphone placed on the table, opened the meeting with the remark, "I hope we will keep up the serious work throughout the entire exercise—especially with such a complex exercise."

The attitude of the participants was indeed very serious. Their reports on events, analysis of the situation, and conversations with each other were as though they were in a real state of emergency. In other words, the exercise was treated as a "real" disastrous event. In fact, sometimes the exercise was perceived as being so real that the exercise's organizers had to remind participants that their communications were only those of an exercise and not a real event.

In another visit to a situation room at the Ministry of Social Affairs on 2 June 2015, during Turning Point 15, I also observed the following moments of confusion between an exercise and a real emergency event:

> An event is received from the "event generator." It reads: "A hostel with people in need is evacuated and they ask to be evacuated to Savyon."
>
> A representative from the (Ministry of Social Affairs) situation room receives a report on the phone. Following this conversation, he tells the other attendees in the room: "We need to evacuate 24 people, staff and inmates."
>
> *Representative 2:* "This [Savyon] is a dangerous area [i.e., it is under rocket fire]. It does not make sense. The quietest area is the south. We will direct them to the south or the Jerusalem area. Let's examine Jerusalem."

Representative 3 opens a document and looks for possible places for evacuation in the Jerusalem area.

Representative 2: "You can split them between different institutions if necessary."

The participants debate where to move the evacuees and how they might be split up.

At some point, a phone call is received. A representative answers: "I do not refer them to Savyon. It makes no sense. We are transferring them to Jerusalem."

Towards the end of the discussion, the representatives from the situation room formulate an order and send instructions to the headquarters situation room, the director-general of the ministry, and the various institutions and organizations that they have decided to contact so that everyone can be prepared and can notify the evacuees.

Someone shouts in the room: "Write clearly: 'exercise—exercise—exercise' [to make it clear that this is only an exercise, and not a real situation]."

Even though the participants were consciously aware that what they were dealing with was only the internal and imagined world of the exercise, there were still moments when the simulated events were so close to reality that they suddenly needed to remind each other that what they were experiencing at the moment was only an exercise. Moreover, the familiarity of the events created for the sake of the exercise not only reinforced the world they were inventing during the exercise's implementation but also affected how they viewed the external world, as we can perhaps see in the following scene from the Ministry of Economy situation room:

I'm at the Ministry of Economy situation room. It is Turning Point 2015, 1 June 2015.

At 4:00 P.M., an update comes in that there has been a power outage in Haifa and the Krayot area. A representative in the situation room tells me that it is difficult to "simulate" a real attack, because if there were a power outage then the only way to imagine it would be to turn off the lights.

Shortly afterwards, a situation evaluation begins. Reports flow into the situation room, and the representatives summarize what they know about what has occurred at this stage of the war: "We are today in the first day of the war. Until now, there has been no emergency order from the Ministry of Defense. We are in a state of war, but it has not yet been declared [by the government]. Do not forget that we are in the first day. There is uncertainty. There is still no information about what is happening. NEMA has implemented an evacuation plan on the northern border for families, which consist of about thirty to forty thousand people, and it is important that the district directors know where the people are evacuated from and where they are to be evacuated to."

A representative from another district talks of three places being hit, and of the need to estimate the damage. Another representative declares: "There is a mess in NEMA's system. There are a lot of events. There is a network mess. They confuse us." He is angry, and continues, "I do not know where I stand. We are in an exercise, but tomorrow it could be a real moment."

On the one hand, the imagined reality of the exercise cannot entirely simulate the real. For example, there is no real power outage during the exercise, even though, according to the scenario narrative, this is the event that the participants have to imagine and to which they have to respond. On the other hand, the participants who deal with the incoming reports hardly distinguish between what they are experiencing in the exercise and what can happen "tomorrow" in reality.

We might say that through the scenario exercise, a fictional world is created. This world engages the imagination of what is knowable but also what seems to be not yet real or not really real. At the same time, this fictional world also blurs imagination and reality in the experience of the participants, who both think and act in the exercise as though it were a real emergency and subsequently view reality in some ways as an extension of the exercise.

Actualization: Unexpectedness and Indetermination

In these exercises, the broad virtual event presented in the scenario narrative is actualized and translated into practice through multiple incidents that have concrete repercussions. Many subevents emerge, each with the potential to develop in unexpected ways depending on the reactions of participants and how those incidents interact with other incidents.

As the following example shows, an electricity shutdown is related to its possible effect on pharmacies' ability to operate and physicians' ability to get to work, thus compromising medical aid to the injured:

> On the morning of 2 June 2015, I visit the situation room of a municipality in central Israel during Turning Point 15, which is now open for the purpose of the exercise. Staff at the room receive reports (by phone and computer) and forward specific information to the office of the municipality's chief executive officer (CEO), where the municipality CEO and main co-

ordinators are stationed. A situation evaluation meeting begins at 11.20 a.m. The heads of teams charged with monitoring the effects of the scenario event on local infrastructure and services present their reports:

CEO: This is a state of emergency, but the system can function this way for a long time. The public should obey the directives of the Home Front Command, and when a siren is heard everyone must enter a protected space. These are not easy times, but we can deal with this threat. The blackouts are a problem, and we need to know how to deal with them.

Engineering team head: There are power crashes; we spoke to the power company and they are aware of the problem and said that the power should be back within four hours.

Population team head: Two people were injured during this morning's rocket attack, one low and one medium in severity. We received a notice from the manpower team and are in touch with the hospitals. A team was sent to the location where the incident took place.

Health team head: There is a shortage of physicians. We turned to the Ministry of Health and they said they would send physicians, especially pediatricians. Additionally, 40 percent of the social workers haven't arrived at work, and 805 elderly caretakers left their patients alone.

The participants know that rocket attacks have occurred in particular areas, and they translate this information into concrete developments, such as power crashes, people injured, and shortages of physicians and social workers. What is interesting about these details, which are reported by participants at the situation rooms of various units to their superiors to create a "situation report," is that they are invented by the participants during the exercise itself. That is, the participating units are notified, for example, about the striking of missiles in their area; from that moment on, they "translate" this information into its concrete meanings: how many people were injured and who exactly, what are the consequences of these injuries, and so on. This multiplicity of incidents turns out to be very specific. That is, it is expressed and described with attention to minor details.

For example, during Turning Point 15 I saw the handwritten note shown in figure 3.2, which elaborated the response to various incidents in one municipality. The note reads, "(1) At 24–26 Yitzhak Rabin St., there is no electricity and generators are not working. *We locate the house committee to investigate why the generator is not working.* (2) In 'Noam House,' generators are not working. I contacted the IEC [Israel Electric Corporation] to see if they can assist with providing a generator, because in Noam there are *people with disabilities.* (3) 8 Prophets St. is without electricity. I contacted the IEC. At the same time, we'll see why there is no electricity."

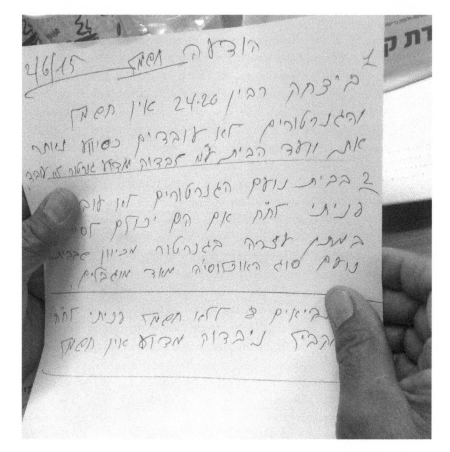

FIGURE 3.2. Turning Point achieves reality through specificity. Photo by the author.

In this short note, for instance, we can see that the details about the incident that were added by the participants in the exercise were very specific (these details were not part of the original scenario narrative, even though that in itself was very detailed). For instance, the exact street number for a particular building was given, and it was mentioned that there were people with disabilities at this address. Moreover, the response of this practicing team to the rocket strike at 24–26 Yitzhak Rabin St. and the subsequent electricity shutdown was to investigate why the generator was not working and to add more details to the story.

I was often told by officials within the Turning Point administration that "every rocket has an address," a comment that underscored the level of specificity and detail that they aimed for in the writing of the scenario narrative. However, as the national scenario event is actualized, many unexpected repercussions are developed by the participants themselves, and these further enrich the contours

FIGURE 3.3.   Evacuation drill: Schoolyard as evacuation center. Photo by the author.

of the scenario narrative and its specific, concrete effects and consequences. This development of the narrative throughout the exercise could also be seen in the evacuation drills that were conducted as part of the exercises (see figure 3.3).

Such evacuation drills were among the events planned for the 2015 scenario. While the broad outline of the drills was written in advance, specific logistics were determined by scenario participants. Several schools in central Israel were chosen to serve as reception centers for evacuees from the north after simulated rocket attacks in that area. On 1 June 2015, the second day of the exercise, I visited one of these schools to watch an evacuation drill. Reception desks, staffed by personnel from different fields, were set up in the schoolyard to address evacuees' varying needs upon their arrival at the center.

At 11:00 a.m., the drill started, and evacuees began to arrive at the reception desks. As the scene became more hectic, the reception staff continued to welcome evacuees with calm, brisk efficiency. An evacuee approached the social work and welfare desk and said that his daughter was claustrophobic. The representative at the desk asked, "Is she diagnosed as claustrophobic? What are her difficulties?" The evacuee described his daughter's situation and the kind of anxiety from which she suffered. The representative (playing the role of psychologist) agreed the girl could stay outside the school building.

Another evacuee approached the public information desk, where a representative greeted him warmly and asked, "Did you arrive alone?" The evacuee responded, "No, I arrived with my wife and four kids. My wife is gluten-sensitive and needs a gluten-free meal." The representative wrote down the request and told him, "If there's a gluten-free meal, they will contact you. Enjoy your meal and have a pleasant stay." Later, this representative explained that he was operating partly according to guidelines he received prior to the exercise but was mainly improvising in the situation, using his common sense.

One Turning Point official described the way in which Turning Point scenarios were enacted and developed in terms of "child events, grandchild events, and great-grandchild events" that emerge from a "parent event." Once the parent event is activated, a multiplicity of actual, "descendant events" follow. These descendant events cannot be fully written in advance, especially when thousands of incidents can be actualized at the same time. As administration official Eyal Barak explained to me,

> How do we produce the general super-scenario? The challenge that we took upon ourselves . . . was to build a scenario that's as realistic as possible, but at the same time—and that's more difficult—completely closed. What do I mean by closed? When a rocket lands in Frishman Street, let's say that the fire service knows it's in Frishman, and the municipality also knows it's in Frishman, and Magen David Adom [the Israeli Red Cross] evacuated people from there so they also know where it was. How do you make sure all these organizations know that a rocket landed in 32 Frishman Street . . . ? Now, if you have 20 rockets like that each day [of the exercise], you can write a story for each of them. But we're talking about 2,000 or 3,000 rockets a day. You can't really write everything top-down.

The planned scenario narrative, the imagined future, is central to the way in which participants think about the unknown future while conducting their actions in the present. Nonetheless, an important element of the practice of the scenario in the exercise is the continual creation of the future-present event as it is being practiced. That is, the premediated future is not a static image; while it is being exercised, the multiple reactions of the participants create new effects—some of them unexpected. Through this process, the scenario is continually re-created in an indeterminate process.

Put differently, the prewritten narrative shifts and changes as it adjusts to issues that emerge only during the actual practice of the exercise. Thus, while overall the exercise is a predesigned world, one of its most important aspects is that unpredictability arises at the level of the local actors and the interactions that take place between them. As a consequence of this logic, participants constantly find themselves both creating and unfolding incidents in ways that involve actors picturing the emergent situation and acting upon it through organizational means and practices. However, since these actions themselves interact with the actions of other actors, the outcomes are also unpredictable and dynamic.

In the case of an earthquake scenario exercise, the process of remaining close to the written narrative can be even more difficult. In this case, the information provided to the trainees is preliminary and very limited. Participants usually get information only about the location of the earthquake and its intensity (magnitude), and then the majority of the story is developed while they practice the exercise. That is, participants both create and respond to the narrative at the same time.

> It is 11 June 2017, the first day of an earthquake exercise. I am in the situation room at Haifa County. The room is staffed with people reporting on incidents in the area in the aftermath of an earthquake.
>
> In this case, the only story given to the participants by the directorate of the exercise is that there has been an earthquake of a certain magnitude at a certain location. All other secondary events—including roads being blocked, hospitals and schools being damaged, people considered missing, etc.—are being made up by the participants themselves in order to translate the big picture into concrete developments. In other words, the participants themselves are transforming the event into concrete incidents that are then distributed to other units as additional information. Through these developed consequences, the participants begin to understand the meanings of the earthquake event and the problems that may arise within their areas of responsibility.
>
> The emergency desk operator enters the operations room and asks everyone to get on the phone with relevant government ministries within ten minutes to obtain a situation assessment. Reserve soldiers are manning the computers in the operations room. The emergency desk officer turns to them and asks them to check an event related to a blockage in the road to Carmel Hospital: "Can they or can they not receive patients?"
>
> The Haifa District emergency manager reports to the people in the room: "There's an update from the police: the earthquake epicenter is Beit She'an. There is also wreckage in Haifa and its surroundings. The police are working on a statement that will be issued to the press. The road to Carmel Hospital is blocked. Many people are leaving their homes for open spaces. Everyone should continue communication with government ministries and local municipalities for further details. . . ."

Concomitantly, another representative turns to one of the reserve soldiers, who has been in touch with government ministries, and asks questions regarding the schools that were damaged in order to get more information.

The Haifa District emergency manager gathers everyone together and says: "We are in the midst of gathering data, trying to understand what happened in our realm—from local municipalities to government ministries. A difficult problem that we need to deal with is that this is the middle of a school day."

An emergency desk operator urges: "Try to verify the reports! In the information flowing into the operations room, there was information about a hospital collapsing, and they said that it was Carmel Hospital, but on the news they said that it was Rambam Hospital. Call the Ministry of Health once again to verify."

In this example, participants located in a situation room (see figure 3.4) are provided with initial and general information about the earthquake and on this basis of this information have to use their imagination to invent secondary events that are highly concrete (e.g., roadblocks, missing persons). These newly invented events are communicated as information to other participants, who then put all the information together to understand the bigger picture.

These developments cannot be devised in advance. They are created by the participating units as part of their translation and perception of the situation. Events received from the central administration (the main scenario) require the participants to react by translating generally reported incidents into concrete effects and diverse possible actions that could be taken in response. Together with the reactions of other actors, these reactions create new situations and challenges that are reported to the central administration to create an aggregated situation evaluation.

Counter-Actualization, Rebuilding the Puzzle

During the actual week of each exercise, Turning Point administrators would convene at least twice a day to compile all the situation reports they received. These meetings were held in a room whose glass door was marked with a "No Entrance" sign. In attendance were all functionaries of the central administration and those in charge of the operation of government offices and local municipalities (though only twenty or so members of the exercise administration actually presented updates at the meetings). They were seated at long desks, their

FIGURE 3.4. Situation room, Turning Point 15. Photo by the author.

places marked by nameplates. Each person presenting would report on the main incidents that had occurred in his or her sector that day and on the actions taken by individuals and units taking part in the exercise in response.

Interestingly, these situation report meetings not only brought together and collated the information that was now being produced by the participants but also could influence the original prewritten narrative, which might be adjusted to take into account developments during the exercise that were reported at these meetings. That is to say, while the implementation of the exercise involved the actualization of multiple incidents, which various participants enacted and responded to by sometimes adding more details and information to the initial scenario outline, this information was subsequently transmitted back to the central exercise administration (who had written the original scenario) so that they could examine how participants had understood, translated, and reacted to these events. The administration would then sometimes need to adjust the events planned for subsequent days of the exercise to take into account new needs or issues—for example, unexpected reactions of participants—identified through this process.

> Exercise administration headquarters, Turning Point 15, Day 2, 10:00 p.m. All relevant administration personnel are present in the situation room to discuss the current day's situation reports. The various representatives summarize what has occurred in their respective areas of oversight:
>
> *Ministry of Public Security representative:* The training forces are responding seriously all over the country.
>
> *Another representative:* Do they manage to get to every rocket?
>
> *Ministry of Public Security representative:* No, not to every rocket. There are a lot of traffic problems. A lot of population problems. The security forces cannot get to all the rockets, and the public is very upset because of this.
>
> *Infrastructure representative:* The issue of the electricity blackouts isn't working well [i.e., isn't being taken seriously]. There were many blackouts since this morning. Fifty-five blackouts. Some of them lasted four to five hours. But only one was treated.
>
> *Government representative:* At 11 a.m., the prime minister rushed to a protected space along with all of Israel. He made a video call to NEMA to get a situation report.
>
> *Medicine representative:* A hospital director in the north called me and asked not to bring in any more injuries. He has high occupancy rates. There's a mission here: how do we evacuate these injuries? I ask that the number of injured in the exercise be reduced. It should be a kind of "reward" for the people training. We need to respond to what the participating units are doing; and if there is a need to change and update the exercise, we should do it.
>
> *Cyber representative:* Every event that is not a rocket is understood as a cyber event. There is too much "background noise" surrounding the "actual" cyber events that I have prepared for the exercise.

During the meeting, then, various officials complain about either insufficient responses (e.g., to the electricity shutdowns) or overresponses (e.g., the translation of every event into a cyber event). Because the reactions of the participants are sometimes unexpected, officials deliberately aim for a dynamic, adaptable administrative approach that allows them to change the central scenario during the exercise.

In another case, given the severity of the rocket attacks in the north, two municipalities ordered the precautionary evacuation of their towns on the second day of the exercise. As a result, the central administration had to change the targets of the next day's rocket attacks to other cities that had not been evacuated.

After the national scenario is actualized, the various units that are participating in the exercise are expected not only to respond to multiple emerging incidents or actualities but also to work toward discerning the broader picture of the event. That is, their situation reports play an important role at these meetings. In addition to reporting on the scenario's enactment, such reports have a further purpose: they are also a venue in which a comprehensive story, one that goes beyond the multiple incidents they describe, is created. Michael Yair used the metaphor of a jigsaw puzzle to describe this task of moving from the actualization of the scenario to the re-creation of a situation report picture, which would provide the baseline for understanding and extracting some of the exercise's conclusions. These are intended to identify further concerns, new problems that should be taken into account in future planning processes:

> When we let our children assemble a jigsaw puzzle, we show them the picture on the box and the pieces of the puzzle. . . . [In the exercise,] they don't have the picture and not all the pieces of the puzzle. They will never have all the pieces. That's the premise . . . sometimes very important pieces will be missing. I mean, in the final analysis, you don't need to produce pieces of a puzzle, but a story. Whoever produces pieces of a puzzle could only provide a partial solution. [They must] create a picture and complete the missing pieces based on experience.
>
> The challenge is to produce a situation report that's close to [scenario] reality. . . . Every single [unit] produces a situation report according to their level. One of the central problems is that people don't know how to define the situation report elements that they need . . . to make decisions. Now, every second, the elements change, because the event itself is dynamic. That's the greatest challenge, that's the greatest uncertainty.
>
> *Author:* If I use your metaphor, not only do we not have all the pieces of the puzzle, the pieces we do have are constantly changing.
>
> *Michael Yair:* Constantly, and [exercise participants] don't understand that they are constantly changing.

There is thus a difference between the original scenario narrative and the emerging situational picture as it is extracted from the multiple incidents that resulted from the practice of the scenario. While the scenario narrative refers to future unknowns that are made plausible through story building, the situational story or picture is extracted not from the future narrative per se but from the multiplicity of incidents that were actualized in the scenario exercise. Hence, while the narrative of the scenario is a virtual event actualized into multiple incidents in the exercise, the situational picture is created through a counteractualization process, whereby the "big picture" is extracted out of multiple incidents and dynamics that occur during the exercise. Moreover, this picture goes beyond the aggregation of the multiple and specific actualities. It should point to or identify a broader problem, the new vision that emerges from the scenario exercise and can reveal something about the unknown future, about possible problems that could happen if a particular future comes about.

Accordingly, during the situation report meetings I attended, participants were explicitly instructed not only to report on specific incidents and responses that took place in the exercise but also to focus on the larger dilemmas those incidents evoked. During one such meeting, the head administrator responded to the police representative's report on the various incidents falling within the responsibility of the police that had taken place that day by asking, "What are the problems?" He added that he was not interested in the incidents particularly or in the correct police response to them. Rather, he was interested in a further step of analysis: how the exercise and the scenario activated new problems or at least caused the police to establish a committee to discuss these new emerging issues. He thus continued by saying, "The fact that the police initiated a committee to examine the jurisdictional status is what's important. I don't care about the conclusions of that committee." The goal of the practicing units is thus to extract problems from multiple actualities and to attempt to understand the big picture beyond these specific incidents. Doing so does not lead them back to the original scenario, what the administration terms "God's vision," but enables them to understand what is significant beyond their local experiences and to decide how to proceed from that understanding.

Hence, producing a situation report from the experience of a multiplicity of actual events is a key phase in the exercise but not an end in itself. As Eyal Barak stressed in this regard, "The situation report is not the purpose. The situation report is a tool for us to know what the problems are for which we need to find solutions; these are the implications of the situation report." Only when a situation report is assembled does understanding begin to emerge of the problems that must be prepared for.

To use Deleuze and Guattari's ([1991] 1994, 159) language, I argue that on the one hand, "the event is actualized or effectuated . . . into a state of affairs," and on the other hand, "it is *counter-effectuated* whenever it is abstracted from states of affairs so as to isolate its concept." This isolation of the concept, the extraction of the problem, can only happen through the exercising of the scenario narrative, where the mode of jurisdiction of the exercise promotes multiple, unexpected, and indeterminate developments that emerge as part of the practice of the exercise: "To double the actualization with a counter-actualization, the identification with a distance, like the true actor and dancer, is to give to the truth of the event the only chance of not being confused with its inevitable actualization. . . . It is, finally, to give us the chance to go farther than we would have believed possible. To the extent that the pure event is each time imprisoned forever in its actualization, counter-actualization liberates it, always for other times" (Deleuze 1969 [1990], 160).

The various participants in the exercise, then, are expected to counteractualize the events presented to them, to extract the broader event, the comprehensive scenario story, from what they have experienced locally as multiple incidents. They are asked by the Turning Point administration not simply to respond to the incidents with which they are confronted but also to conceptualize and recognize emerging problems beyond specific manifestations.

Hence, once actualized, the well-written scenario, with its hour-by-hour resolution, creates new incidents and unexpected reactions that affect the development of the original scenario. Moreover, in putting together a situation report on the basis of the specific incidents they encounter, the participating units are not asked to merely replicate the scenario event created by the administration: the practicing units are expected to reveal the broader events beyond the specific incidents presented to them, to extract the larger picture from multiple actualities, and to pose new possible problems.

Accordingly, as an uncertainty-based technology, the scenario thus not only conceptualizes the future in a way that is different from that enabled by risk-based, actuarial reason but also enables a distinct form of action. This form of governing is not intended to design a future of controlled specified possibilities. On the contrary, generating the unexpected is part of the process of practicing and developing the scenario, and in addition contributes to the constructing of problems related to the unknown future. Hence, the scenario as an uncertainty-based technology narrates imaginable futures and puts them into practice to proliferate the unexpected through such practice. In this technology, then, uncertainty is not just the basis of a mode of knowledge making about the future but is also expressed in the practicing of that future.

SUBJECTIVATION
Embracing Uncertainty

The first World Energy Council scenario-planning workshop I attended was held in the convention center of a Paris hotel in September 2018. This workshop launched a yearlong process of regional scenario work (as part of the design and implementation of the global scenarios). Four large round tables were placed in the middle of the room in which the workshop took place, each with ten to twelve chairs around it. The scenario-planning team sat to one side of the room behind a separate long table. At each of the four round tables sat a facilitator from the World Energy Council scenario-planning team, who would manage the discussions around that table over the next couple of days. Each participant who entered the room received a badge that displayed his or her name and was directed to one of the four round tables.

Interestingly, during the opening remarks of the meeting, the facilitator explained that the aim was to "provide . . . a framework to think about the uncertain future." That is to say, in the long process of building the narrative of the scenarios and all the attention given to the specific informational content of each scenario, the structure or framework that would be used to think about the future would be a key goal of the process. Put differently, beyond knowledge making (veridiction) about the unknown future, the methodology being used to develop these scenarios also entailed a particular form of power and action (jurisdiction). Furthermore, as I will show below, providing leaders with this form of thinking and planning for the future was also a central goal within the agenda of the workshops.

Unlike the Turning Point scenarios discussed in chapters 2 and 3, which were mainly constructed by a central administrative group and where participants were involved in practicing a scenario that had been developed in advance for them, in the current case the scenario narratives were not framed or built by a central team alone. Rather, they were the outcome of a series of workshops held around the world, whereby members of the World Energy Council from different countries played a central role in the process of building the (global, regional, and sectoral) scenario narratives.

In this chapter, through the case of scenario making at the World Energy Council, I explore how in addition to new modes of veridiction and jurisdiction, scenario thinking and practicing generate a new mode of subjectivation; how these three modes are mutually connected and constituted; and how they together express a form of governance based on a rationality of uncertainty.

World Energy Council Scenarios

The World Energy Council's scenario planners create global energy scenarios that represent alternative plausible futures for the energy sector through a triennial cyclical process that involves various workshops with numerous representatives of central organizations in the energy field, including private companies, governmental ministries, and research institutions from around the world. In 2016, following a three-year process that included fourteen workshops, the council published three future global scenarios with a long-term horizon of 2060—entitled "Unfinished Symphony," "Modern Jazz," and "Hard Rock"— which aimed at informing discussion on the transition to a low-carbon economy. These scenarios, it was claimed, would provide "energy leaders with an open, transparent, and inclusive framework to think about the uncertain future, and thus assist in the shaping of the choices they make" (World Energy Council 2016, 2).

In the "Modern Jazz" scenario, it was envisaged that new opportunities in the marketplace and demands for greater economic growth would lead to increasing deregulation, particularly in Europe. The "Unfinished Symphony" scenario, in contrast, described a situation in which Europeans felt that the project of increasing European unification would lead to a better society for all, with this increased integration strongly influencing the direction energy businesses would take. In turn, the world envisaged in the "Hard Rock" scenario was firmly shaped by questions of national identity, with businesses in various parts of Europe developing in different and sometimes contradictory ways.

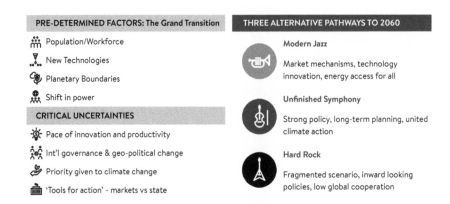

FIGURE 4.1. Recap of the archetypal World Energy Scenarios 2016. Used by permission of the World Energy Council (2019, 5).

As summarized in figure 4.1, the three scenarios capture different futures resulting from a combination of predetermined elements (e.g., a shift in power toward Asia) and critical uncertainties (e.g., whether priority would be given to climate change) related to state policies and the market. Modern Jazz represents a more successful market, Unfinished Symphony represents more successful state policies, and Hard Rock represents a low degree of success in both.

The three scenarios thus set out three plausible—and challenging—stories about possible developments in the energy sector up to 2060. They were designed to facilitate thinking about how various business strategies and government policies might play out over key developments related to four particular trends that had been identified as key factors in how the future might unfold: (1) greatly reduced levels of growth in population and the global labor force; (2) the introduction of new, powerful technologies; (3) greater awareness and respect for the planet's environmental boundaries; and (4) a shift in economic and geopolitical power toward Asia.

In terms of energy, each scenario exhibits different, and highly concrete, outcomes. In the Modern Jazz scenario, by 2060 the world has low-carbon, resilient energy systems (World Energy Council 2016, 30). In this scenario, the emergence of innovative and disruptive technologies results in highly diverse energy sources. For instance, increasing utilization of natural gas, biofuels, and electric vehicles fuels a shift to a diverse mix of energy sources for transportation. In addition, Modern Jazz involves soft policy interventions on local, national, and regional levels (rather than at the international level) that lead to slower growth in carbon pricing and taxation on the one hand but to rapid improvements in technological solutions for renewable energy and energy storage on the other. Thus, in the Modern Jazz scenario, energy becomes more diverse, available, and accessible. Nevertheless, and although carbon emissions are drastically reduced owing to these developments, the target of reducing emissions to a level of 1,000 gigatonnes of CO_2 per annum is not achieved (World Energy Council 2016, 34).

In the Unfinished Symphony scenario, 2060 sees a world that has a low-carbon, resilient, and integrated energy system (World Energy Council 2016, 31). In this scenario, strong policy interventions, regulations, and cross-sectoral standardization on national and international levels (including, for example, green subsidies) together with a declining rate of demand for energy, large-scale technological innovations, and integrated solutions result in improved efficiency and reduced carbon emissions. However, these developments also lead to the collapse of many large utility companies. Consequently, in the Unfinished Symphony scenario, there is a "global energy system" that shifts from using fossil fuels to electricity (World Energy Council 2016, 53), and by 2060 the world is close to achieving the target figure of 1,000 gigatonnes of CO_2 per annum (World Energy Council 2016, 61).

In the Hard Rock scenario, by 2060 energy systems suffer from underinvestment and weak resilience. The world becomes ever more fractured as national interests outweigh global collaboration and action, which are necessary for addressing climate change (World Energy Council 2016, 31). In this scenario, energy solutions come to be driven by security needs, limited economic growth, and lack of international cooperation, and governments create energy policies that favor nuclear energy, as well as cheap and domestically available energy sources (e.g., solar and wind) (World Energy Council 2016, 73). In the Hard Rock scenario, then, climate change is a lower priority, surpassed by security and economic issues. Consequently, the target of 1,000 gigatonnes of CO_2 is exceeded, global temperatures rise by (approximately) 3°C, and climate change causes physical and economic destruction (World Energy Council 2016, 80).

In 2017, a year after the production of these 2060 scenarios, the World Energy Council scenario planners revisited the scenarios and decided to retain them as a framework for the next cycle but with a focus on the future up to 2040. The World Energy Council thus began the process of updating the scenarios to take account of recent developments and to focus specifically on the period leading up to 2040. The new focus would be on an "innovation twist to 2040" and the use of scenarios to explore new exponential growth opportunities that might be involved in speeding up the process of energy transition in an era of disruptive innovation. Accordingly, the World Energy Council scenario-planning team conducted a series of regional scenario-planning workshops in different parts of the world in 2018–2019, seeking to capture different perspectives on the three initial scenarios and how they might potentially unfold in diverse cultural and geopolitical contexts. This process began with experts interviewing energy leaders. These interviews were then followed by several workshops focused on framing the energy problems facing the contemporary world and then, finally, the building of the revised scenarios as a global multiregional model.

In terms of knowledge production, the stories contained in the World Energy Council scenarios confront the problem of the uncertain future neither by making predictions nor by attempting to provide accurate assessments based on past information. Rather, they are plausible stories aimed at provoking a future-now thinking, through which participants might think about possible drivers of the uncertain future and experience various future energy trends. One way in which this works, as scenario expert Nick Mayer explained to me, is that when people build scenarios they are encouraged to search for and identify "turning points." Having thought through three different scenarios and their unique turning points, when these people look at the world around them they are able to monitor signals that help them better understand which future scenario we might be heading toward. Another way in which this works, he further explained, is that predetermined elements that are used for building scenarios project long-term trends that can lead to significant changes that need to be considered.

In either case, the process does not set out to create ultimate knowledge about (possible) energy futures. Rather, the process represents the creation of a tool through which subjects can think about and act on the future. The representatives of the various energy organizations that take part in the scenario-building events are first trained to accept uncertainty as a preliminary standpoint as they are confronted with their own assumptions about the future and then practice a process that involves narrative-building techniques, storytelling capabilities, systems thinking, and quantified illustrations, which together are used to create a framework for thinking about and addressing the future.

Critical Uncertainties and Plausibility

Throughout the process of updating the 2060 scenarios to highlight the period until 2040, the focus was on key drivers of change: "what is inevitable and what is uncertain in future developments toward 2040." The goal was thus not just to update the scenarios with regard to new information that had come to light since the conclusion of the previous scenario-planning process in 2016 but also to identify new uncertainties and new domains in which unpredictable paths could be developed in the future. Nick Mayer, in his role as one of the facilitators, explained in his opening remarks: "We *cannot banish uncertainty*, but we can offer to stimulate thinking on what might be certain and uncertain, and where new opportunities and risk exposures might be" (emphasis added).

Hence, we see that the preliminary standpoint adopted by the facilitator is one of acceptance of the potential uncertainty of the future—a mode of subjectivity that represents a shift away from modes that seek to reduce uncertainty or cancel it out completely. In accordance with the former mode, the scenario narratives were to be built around key uncertainties in the field of energy and more broadly "in the world," through a dynamic process in which the narratives would change and shift in response to these uncertainties.

During the first session of the Paris workshop, participants were divided into groups and introduced to predetermined elements (things that were supposedly already known) and key uncertainties that needed to be discovered and discussed in relation to each of the three scenarios.[1] In addition, participants were specifically asked to think about *critical uncertainties* related to each of the five following themes that would be used to assemble the scenarios: (1) international government and geopolitical developments, (2) the pace of innovation, (3) the priority of environmental issues, (4) management of the energy sector, and (5) infrastructure and investment.

It was acknowledged that shifts in international governance and geopolitical changes could critically affect energy futures (see figure 4.2). In the 2016 Modern Jazz scenario (World Energy Council 2016, 39), for example, global social-economic interconnectedness and complexity mean that regions with "less mature" political systems and regions with "more mature" political systems eventually have to cooperate. Thus, international and national institutions, governments, and industrial and business actors from across the world work together to promote innovative technologies and partnerships to develop clean energy solutions.

As automation becomes ever more widespread and the "low-skilled labor" workforce (i.e., nonexperts who can often be replaced by machinery) grows, the pace of technological innovation and productivity growth shape energy futures. In the 2016 Modern Jazz scenario (World Energy Council 2016, 36–38), global economic competition and growth (encouraged by free trade and migration) and industrial changes enabled by digital technologies lead to sustained beneficial productivity and technological innovation. The combination of economic prosperity, highly skilled labor, and technological innovation, in turn, allows more efficiency in managing carbon intensity and increases resilience (e.g., through the use of data and algorithms to manage risks and anticipate failures). Meanwhile, in the 2016 Unfinished Symphony scenario (World Energy Council 2016, 56–58), policy makers in both developing and developed economies balance economic

FIGURE 4.2. Thinking on the pace of innovation to assemble the three scenarios, Berlin workshop, 2018. Photo by the author.

growth with long-term development considerations and, as part of this, invest in digitalization of the economy and "intelligent infrastructure," which leads to economic efficiency and productivity (e.g., through the creation of "smart grids" and "smart cities" that use real-time communication and analysis to reduce maintenance costs). These developments, in turn, entail higher efficiency and productivity in both the public and private sectors, thus substantially lowering marginal costs, including those of energy.

The priority that publics and their governments give to climate change and environmental issues is another critical uncertainty that affects energy futures. For example, in the 2016 Unfinished Symphony scenario (World Energy Council 2016, 61), local and national support for policies that focus on energy efficiency and long-term planning, as well as globally unified action through cooperation and a strong global system, result in reduced greenhouse gas and carbon emissions.

The preferred way for managing the energy sector, whether with more emphasis on the state or on the market, varies considerably in each scenario. In the 2016 Modern Jazz scenario (World Energy Council 2016, 34), the market dominates as private industry becomes the strongest actor, and policy makers only react with soft policy interventions to address externalities. In the 2016 Unfinished Symphony scenario (World Energy Council 2016, 53), however, global and state institutions are more dominant than industry forces, and thus strong policies are created to shape markets while businesses adapt to comply with the new environmental and social standards that are set. Finally, in the 2016 Hard Rock scenario (World Energy Council 2016, 73), national governments are the dominant force, with state-owned enterprises turning into a widespread tool. Thus, both markets and policies become fragmented as they serve local or national ends. Security then becomes the main driver in the future of energy.

Throughout the regional scenario work process, participants were asked to think about how each of these themes might affect the three scenario narratives, along with different developments that might be expected in each scenario. Thus, the workshop started with a focus on rethinking and updating the previously created scenario narratives, while taking into account various possible developments in the future. It was about updating the information and content contained in these narratives, and making changes that could point to more nuanced and up-to-date stories. These future stories, however, should not be confused with predictions.

On one occasion, a participant asked, "What is our target scenario [for the energy world]?" The facilitator answered that they were not providing a target-based scenario (i.e., normative scenarios) and added, "Our scenarios are not about what we *want* to get in the future, but about which possibilities the future opens up for us" (Lorie Taylor, facilitator and scenario expert, Paris, 2018). Later on, Nick Mayer explained to me that participants were often reminded, "It is [only] a scenario. It doesn't have to happen. It's not supposed to happen. . . . Our goal is to make the scenario plausible, not possible."

Hence, the specificity, concreteness, and reasonableness of the scenario narratives were not about the taming of uncertainty. Rather, they were utilized to assist in the creation of a good story that always, inherently, remained plausible (but not a possibility, in terms of past-based events) and subject to change—that is, to reexpress the idea that the potential future could always unfold in a different way. Thus, during one of the workshops, a participant asked, "Why are we describing the future in terms of what should have happened already in the 90s [after the fall of the USSR]. And it didn't happen. If it didn't happen then, why would it happen now?" One of the World Energy Council scenario-planning experts immediately responded by saying that scenarios were not predictions or possibilities, only *plausible* stories. Hence, the stories they were developing were not expected to be realized in the future (although they might). Moreover, instead of consisting of "memories" of *past* (already known) events, participants were told frequently that the scenario narrative should be "memorable." As workshop facilitator Lorie Taylor put it, "[Good future-oriented frameworks] have to have memorability. . . . We are still refining those scenarios so they are accessible to people and they can use them intuitively; so that they can adopt the frame, they can situate themselves in that future, and they can explore it for themselves. For me, a good set of scenarios is like a theater stage that you can stand on and act out your future. They are not *War and Peace* chronicles to be memorized. They are memorable, not memories."

Hence, uncertainty rationality, expressed in the narrative building of the scenarios, is also linked to a unique temporality. Participants need to think about what comes from the future and imagine that they are living in this "future present" and create "memories" of things that are yet to emerge. This is not to say that in creating "present futures" they realize the future yet to come. Rather, doing this is a way of making the scenario stories *realistic*. As Lorie Taylor explained to me during one of the workshops, "memories are pulled from the future; we are implementing [in the present story] what is imaginable," and this implementation provides the sense of realism. Language is important for this process. As Nick Mayer told me, the way people perceive the future is usually conditioned by a cultural context that enables or restricts discourse about the future. Ambiguous or abstract language avoids the pitfalls of a strong but singular view of the future that exists in some cultural contexts.

Moreover, once a future story has been created, a path from the future to the present can also be drawn. This idea could be seen in the request, repeated at almost every workshop, that participants bring to the group's first meeting, usually for the opening session, a "signal" from our present time that seemed to be connected to one of the various scenarios under discussion. The issue here was whether the participants could discern any signals in the present that might be linked to a particular future, one of the three scenario narratives. When the meetings started, participants were asked to present the signals they had brought. At one workshop, a few participants referred to the *gilets jaunes* or "yellow jackets" as their signal. This was during a period of strikes and political protest in France (2019), and many of the attendees at the workshop had witnessed these events and thought they might be a possible signal of the beginning of a Hard Rock scenario. Hence the process of creating future scenarios was also involved in re-creating pathways from the future to the present. That is to say, through having a preliminary vision of the future, a story of a plausible Hard Rock scenario of the future of 2040, people began to interpret current trends according to such images. In other words, the temporality of narrative building shifted back and forth between imagined uncertain futures in the present—that is, "future presents" (Luhmann 1998)—and the effect of the "future anterior" (Derrida 1988). In other words, once particular futures had been created, they re-effected the present through a process of premediation (de Goede 2008a; Grove 2012), as I will discuss further in chapter 6.

Narratives, Language, and Storytelling

The process of creating the updated scenario narratives took place over almost a year and involved numerous preliminary meetings, group workshops, and the production of various documents. During these events, it was often stated that participants were involved in *narrative-based* scenario planning—a reference to the importance of the story that they were collaboratively building throughout the various meetings. The same idea was also expressed in the fact that in addition to the scenario-planning experts, a professor of literature attended the workshops and helped to facilitate the process of writing the narrative for each scenario. "Scenario building," she often said, "is an education exercise for the future, and the future is always and only *a story—a fiction*" (emphasis added).

Throughout the workshops, participants were also guided on what good storytelling requires and how to create an effective and useful story line. To assist them in creating their narratives, it was explained to them that a story line represents a series of events and how these events are linked in time and in relation to different actors. Every story has a start, a middle, and an end, they were told (see figure 4.3). It sets out what happens and how and thus connects the future to the present. The middle part of the narrative is where the turning point usually occurs, bringing the participants to think of a world that is substantially different from the one that might be envisaged on the basis of forecasts. Narratives should reflect the drivers of change: the predetermined facts and the critical uncertainties. A plausible story line answers the question what might happen rather than describing what we would like to happen. Events are developed into other events or impacts that create more events, and this process is extended gradually into the future (2040). The narrative would be developed from a series of events and impacts.

During one workshop (Paris, 2018), following a discussion of the broad framework of the three scenarios, participants were asked to take two specific "future drivers" and develop them into a detailed story composed of multiple events, describing the concrete results or impacts brought about by those events. Seated around tables in groups, participants were encouraged to write down different events for the three different future scenarios on sticky notes. Colored blue, orange, and yellow, these notes were then posted on the wall in areas marked for the relevant scenario archetype (blue for "Unfinished Symphony," orange for

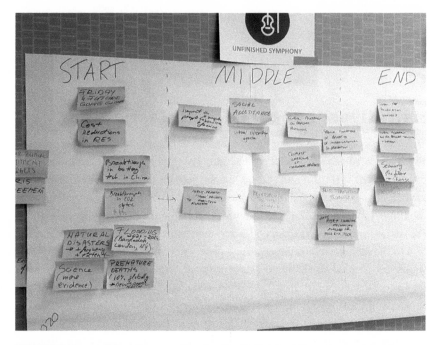

FIGURE 4.3.   Building the narrative for the Unfinished Symphony scenario, London workshop, 2019. Photo by the author.

"Modern Jazz," yellow for "Hard Rock"; see figure 4.4). The result of this part of the workshop was that next to each table was a board with three scenarios, with the title of each scenario being accompanied by various sticky notes that described events that had "happened" on different dates and in different places and had been caused by different developments (e.g., global warming or political change).

On the face of it, there could be an almost unlimited number of possible future events, and each one of these events could be relevant to more than one scenario. The individual stories produced were thus highly contextual and depended on the participants in the workshop, the particular makeup of each group, the interactions of its members, and individual emphasis. I joined one of the groups working on the Hard Rock scenario. We discussed the theme of innovation in relation to this scenario and began to outline some geopolitical events, with specific details, as requested: War in sub-Sahara; Donald Trump is re-elected in 2020, and again in 2024 (the law was changed so that he could be elected three times); OPEC collapses and is dissolved in January 2020.

Providing the background for this discussion was a draft document based on regionally focused elaborations of the 2016 World Energy Council scenarios,

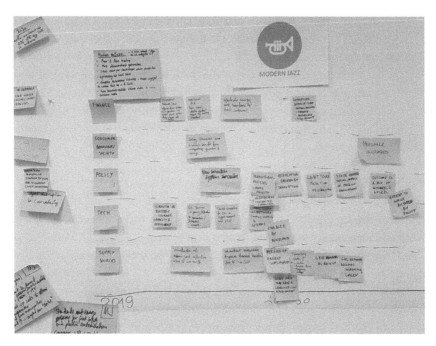

FIGURE 4.4.   Sticky notes with events placed on the Modern Jazz timeline, London workshop, 2019. Photo by the author.

which had been created by a group of World Energy Council members in 2018. Within this document, for example, it was suggested,

> In *Hard Rock,* EU citizens become more and more dissatisfied with suffering from heavy-handed top-down bureaucracy on the one hand and being ravaged by global market forces on the other. Right-wing, nationalistic candidates win surprising victories in many countries, and insist on going their own way in an increasingly *#mefirst* world of Brexit, "America First," growing pressure from China, and a disruption of the global trading system. The ties that hold the Union together begin to fray, and the result is a surprising growth of inequality both among and within countries. Individuals revert to their own identities and cultures, which in many cases are not geographic. (World Energy Council 2018, 8)

These generally regional developments were connected to the more global geopolitical events that had been mentioned in the 2016 Hard Rock scenario:

> The US and EUR initially focus on addressing instability in the Middle East and stemming the flow of migrants, but this proves challenging. . . . Civil wars last through 2020, and continue to spur terrorist acts and

mass migration with global impacts. . . . Just after 2030, China surpasses the US to become the number-one economy and Asian markets, once the destination for over 25% of US exports, are increasingly dominated by Chinese products. China's rise leads to increased geopolitical tensions. This creates additional incentives for the US to create a balancing coalition to contain China by strengthening its ties with EUR and Latin America and building alliances with China's neighbours, Japan, India, and South Korea. (World Energy Council 2016, 79)

We then began thinking of events that might be connected specifically to energy, not just broader geopolitical issues. For example: What would happen to the Nord Stream pipeline in 2020? Would countries without fossil energy use wind and solar power to improve their energy systems? And so on. Answers to these questions had to be contextualized within the concrete scenario, as evident in the 2016 Hard Rock scenario: "As international energy trade declines, widely fluctuating oil and gas commodity prices persuade national governments to look to renewable energy as a domestic source of electricity generation. Solar, wind and geothermal generation grow more rapidly than any other fuel source in primary energy to 2060, averaging 4.2 percent p.a. in the period" (World Energy Council 2016, 89).

During the next stage of the workshop, each group took all of the sticky notes it had created for each of the three scenarios and combined them with those produced by the other groups for the same scenarios. The posting boards were now reorganized, with each group working on a single scenario narrative that included all of the events suggested by all of the groups in relation to a particular scenario (two groups worked Hard Rock scenarios). The idea was to have as many events as possible for each scenario and to use these to begin drawing up an organized and detailed narrative for the future (according to the Unfinished Symphony, Modern Jazz, and Hard Rock scenarios). Hence, the initial broad framework of the three scenarios at each table was translated into many specific events with concrete locations, times, causes, and effects. As the facilitators would put it, "You need to translate the narrative into specific events that address the questions of When? What? Who? If there are no (specific) events, we cannot tell a story."

At various times throughout the workshops I attended, members of the World Energy Council team would describe scenario planning as a craft, arguing that the more one practiced the technique, the more it would become a way of thinking and a shared new language among both experts and users: "If you want to bring about a paradigmatic change [in terms of how one thinks about and prepares for the unknown future], you have to create a new language" (Lorie Taylor, Berlin, December 2018). Accordingly, particular emphasis was placed on the creation of a new language in the process, as Nick Mayer explained during a conversation: "You first have a framework [i.e., the three scenarios], then interviews and narratives, and then concepts and terms. You cannot work without the concepts. . . . The workshop is a way to figure out a new language. It is especially exciting when it happens in these events [i.e., when people start using this language]" (Day One, Paris, 2018).

Indeed, at this point of the process, one could see the embodiment of the "new language" among participants and correspondingly how their entire perception of the future leaned toward the embracement of uncertainty, dynamism, and nonpredictability. In this regard, it should also be noted that the workshops were conducted in a very dynamic fashion. People would stand up and walk between the various boards, while individual notes would pass from one board to another, shifting location and changing the order of events. That is, both content and form were continually shifting and developing throughout the workshops.

For the next stage, participants were asked to link the multiple events they had suggested into a more comprehensive picture and to create the scenario narrative. After all of the events created separately by each group had been gathered together and divided up according to the three scenarios to which they related, participants began to create coherent narratives for each of the three scenarios. They began to narrate the integrated future story in a chronological manner, from the present to the future. At first, they considered significant change factors in relation to politics and technology. For example, they discussed improvements in battery technology and the end of storage problems, as well as the notions that the domestic market was developing, there was less governmental power and more local participation, but inequality was growing at the same time. As part of this process, they reorganized the events that had been attached to the board in accordance with the narrative they created, using them to tell a story of future events from 2019 to 2040 (see figure 4.5).

Significantly, Lorie Taylor emphasized how important it was that participants explicitly practice storytelling. Beyond simply creating and inventing the content for specific stories for each scenario, they were asked to practice the method and told, "We are all going to learn to be good storytellers" (Berlin, 2019). All participants would practice telling stories, and each group would choose one person to represent it, to present the story that the group had created. In addition to building the scenario story, then, these workshops were also about getting participants to *practice their storytelling capabilities*—and thus to embrace not only the content of the scenario-narrative but also the way of thinking and practicing these future stories.

Indeed, that participants would practice to become storytellers was a key goal of the scenario-planning process and was intended to create a sense of ownership

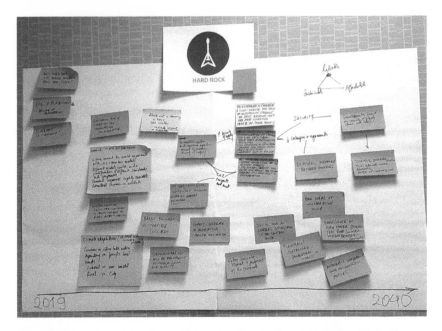

FIGURE 4.5.   Organizing the events according to chronological order for the Hard Rock scenario, London workshop, 2019. Photo by the author.

(both of the created stories and of the method) among them. Through these inductive, reflexive, creative practices and processes, the scenario-planning team sought to ensure that the participants would accept not just the framework of the scenario narratives but also the process through which they were created—that is, the experience of working with uncertainty and thinking about the future through plausible (completed and dynamic) scenarios.

Becoming storytellers, participants were instructed to take an active role and to stand by the board in front of the group and tell a story in one of two possible ways: either they could recount their story narrative from the present (2019) to the future (2040) and show a chronological development of events into the future or they could tell the story from the future back to the present, as though looking back from 2040. As Nick Mayer commented, "We encourage you to think not just about your experience from the past, but also of what comes from the future, as historians of the future." In another workshop, during the stage when participants were telling the stories that they had developed, one of the participants asked, "Are we now doing forecasting?" Nick Mayer answered, "Imagine that we are [now] living in this world. If it were the world that we live in now, what can you say about technological innovation in it?" (Tallinn, 2019). Put differently, these storytelling practices were embedded in a particular temporality of future-now thinking. Lorie Taylor further explained: "So what we are trying to build in the World Energy Council scenarios is a flexible toolkit that allows us to work with different types of futures thinking, visionaries, performance space projections . . . and the issues monitored, which are like attention tools."

In this context, Nick Mayer commented that "the main thing is to create stories/narratives and storytelling in an analytical and methodical way about possible futures. That's the main thing—that's the goal of the process." Thus, as previously noted, through this type of involvement and participation in creating the scenarios, the goal was to encourage participants to embrace the method and to develop a sense of ownership toward the scenarios they had themselves created. This would then lead them to implement the scenarios in their organizations as a tool and a framework to assist with planning for the future.

Modeling and Illustrating

The scenario-planning approach used at the World Energy Council is mainly a qualitative one. This was emphasized many times throughout the workshops, often by way of contrast to other approaches based on quantitative models that are common within scenario planning, which participants might thus have expected to encounter during the workshops. Nevertheless, in the final stage of the scenario-building process, quantitative models were also used by the World Energy Council scenario-planning team.

During a conversation I had with Nick Mayer over one of the lunch breaks (Berlin, 2019), he used a metaphor of the brain to explain the differences between qualitative and quantitative models. Showing me the Modern Jazz team board, he explained how in the workshop a new (local) narrative emerged, one that the World Energy Council's scenario experts could not and did not want to write on their own. Rather, their goal was to get participants from various countries to practice the scenario story so that they would take the new stories, which they had written themselves, back to their countries and organizations and work with them there. Nick further explained to me that the World Energy Council scenario-planning team did use quantitative models in their longer scenario framework but only at a later stage in the process. In terms of how they were viewed and used by the scenario-planning team, he emphasized that these quantitative models were only an illustration of the narrative, not information in itself or a model of the future. "Making quantitative models," he commented toward the end of our conversation, "activates other areas of the brain: narrative-building activates areas of *imagination*, and the [quantitative] models activate areas of *logic*." Interestingly, this biological explanation highlights how the use of scenarios involves not only different knowledge and power systems but also a different mode of subjectivation. Indeed, as I show below, quantitative models were perceived by some participants as contradicting the mode of embracing uncertainty and the narrative forms of knowledge making learned thus far in these workshops.

The qualitative models used in the workshops were first introduced to me during an interview with Nick Mayer at the World Energy Council's head office. During our meeting, Nick handed me the booklet that his team had produced to present their work on the World Energy Council 2016 scenarios framework. On opening it to briefly skim through its contents, I noticed that each of the three scenarios (Unfinished Symphony, Modern Jazz, and Hard Rock) was presented with a long descriptive text on one side of a page and quantitative information, displayed in graphs and charts, on the other. I immediately asked why such models had been included in what I understood to be a "narrative-based methodology." Nick replied, "If I want [managers] to read *this* [pointing to the written text about the scenarios], I also have to present *that* [pointing to the graphs]." As managers are used to receiving information in the form of numbers and graphs, the World Energy Council's scenario-planning experts had to translate their narratives into such forms to convince managers to use the scenario model. This also meant, as Nick explained in a subsequent conversation, that the quantitative models must be highly credible and produced by "world-class" experts.

The same issue was also raised during an interview with Lorie Taylor:

> AUTHOR: So why do you use models?
>
> LORIE TAYLOR: For consistency. And, also, it tells people . . . [it is useful] as long as they use the numbers illustratively, rather than predictively. You have audiences that need the numbers. However, [in the way that we use models as illustrations to qualitative scenarios] no one can read the numbers and tell you its meaning without the story in which the number is in.

At the same time, Nick Mayer explained that the problem with the quantitative models used by other experts is that they are too "heavy"—it is difficult to modify them when new information arrives and the future turns out differently from what had been anticipated. In his view, people were often too busy looking at the model instead of reality, trying to shape reality into what was set out in the model, not the other way around. Or, as another scenario-planning expert put it to me, "They build the model based on their experience, and when their experience doesn't fit the model, they stop looking at experience."

Therefore, in the World Energy Council's scenario-planning activities, models were not used for the sake of transforming qualitative information into better assessments of what was possible or probable. Rather, such models were used only as an *illustration* of the qualitative narrative scenarios. Hence, the emphasis was on the process of the narrative making, which made room for dynamic developments and flexibility, rather than on a fixed image provided by a qualitative model. However, after training participants to work with narrative scenarios, introducing the quantitative models to the workshop participants was not an easy task. In one of the workshops (Tallinn, 2019), the last session was dedicated to presenting these models. The facilitator, energy expert Terry Striker, took the key assumptions from the narrative as input for the model, explained that he had converted the scenario narratives into models and diagrams, and suggested that the participants look at the models to see what had happened to the themes and events they had worked on. They were thus checking whether the model matched their expectations of what it should look like.

There was a certain degree of resistance, and perhaps even disappointment, in the audience as the models were presented to them. The participants were unable to clearly identify what challenges emerged from the stories because the models lacked range. Thus, the immediate reaction of the participants was that the qualitative models did not do justice to the long and in-depth process of narrative building in which they had engaged. They were seen as insufficiently rich in comparison with the narratives. The differences between the three scenarios suddenly appeared very small, and it was sometimes difficult to distinguish between them. At this point, then, the process successfully shifted the participants' assumptions (particularly around the turning points).

For instance, in the 2016 report, the three scenarios express different changes in the share of fuels used in transportation (World Energy Council 2016). These quantitative changes can be seen in the charts shown in figures 4.6a–4.6c. In 2014, the use of electricity was 1 percent of the total share, while oil constituted 92 percent of the total share. In the Modern Jazz scenario, by 2030 electricity use rises to 2 percent, while oil use decreases to 90 percent; and by 2060, electricity use rises to 8 percent, and oil use decreases to 67 percent. In the Unfinished Symphony scenario, by 2030 electricity use similarly rises to 2 percent, while oil use decreases to 87 percent; by 2060, electricity use rises to 10 percent, and oil use decreases to 60 percent. In the Hard Rock scenario, by 2030 electricity use also rises to 2 percent, while oil use decreases to 90 percent; and by 2060, electricity use rises to 4 percent, while oil use decreases to 78 percent.

Ideally, as Nick Mayer explained, the quantitative and qualitative parts would be fully aligned. Nevertheless, in the quantitative presentations of the scenarios, the changes in electricity use for transportation between 2014 and 2030 seem both insignificant and almost identical in the three scenarios. Qualitatively, however, there is much more to the matter. In the Modern Jazz scenario, "EVs [electric

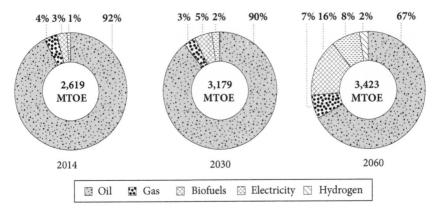

FIGURE 4.6A. Changes in share of fuels used in transport within the Modern Jazz scenario. Used by permission of the World Energy Council (2016, 44).

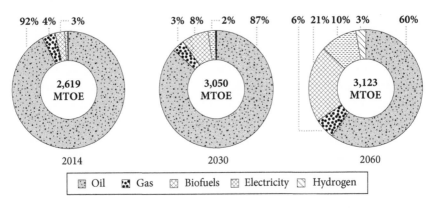

FIGURE 4.6B. Changes in share of fuels used in transport within the Unfinished Symphony scenario. Used by permission of the World Energy Council (2016, 65).

vehicles], supported by technology breakthroughs in battery technology, growing availability of distributed systems that accommodate EV charging, and continued penetration of pure-play EV manufacturers like Tesla that stimulate infrastructure projects, underpin the growth of electricity in transport. Traditional manufacturers such as Ford, GM, BMW, Nissan Volkswagen and Toyota continue to evolve their product offerings to remain competitive. By 2030, there are more than 63mn EV and hybrid EV globally" (World Energy Council 2016, 44).

In this instance, factors such as improved battery technologies, distributed systems, and the commercial interests of large car manufacturers catalyze the growth in the production and use of electric vehicles, which entails a rise in the usage of electricity in transport.

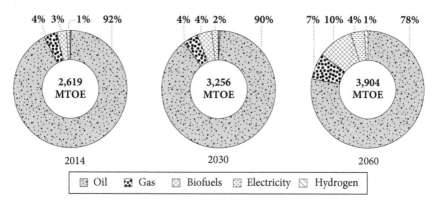

FIGURE 4.6C. Changes in share of fuels used in transport within the Hard Rock scenario. Used by permission of the World Energy Council (2016, 83).

In the Unfinished Symphony scenario, "Policy mandates substantially impact consumer demand for light-duty vehicles: Car-ownership grows at a moderated pace, and the mix of car technologies is highly diverse, including biofuels, EVs, and natural gas.... Through strategic planning, emissions standards, and the build-out of smart infrastructure, the electrification of transport continues to build momentum. By 2030, more than 86mn EV and hybrid EV are on the road" (World Energy Council 2016, 64–65).

In this case, policy interventions, investments in infrastructure, strategic planning, and technological developments incentivize the demand for diverse energy sources, including electricity. This leads to a rise in the use of electric vehicles and hybrid electric vehicles.

In the Hard Rock scenario,

> With reduced capacity for infrastructure spend, demand for personal transport remains high to 2060. The number of cars in the world grows 2.6 times in the period, reaching 1.5bn cars in 2030 and 2.9bn cars in 2060. Reduced capacity for infrastructure build-out and lower economic growth means transport fuels are slow to diversify. Demand for petroleum-based transport fuels grows at a pace of 1.2% p.a. to 2030, dropping to 0.1% p.a. beyond 2030 as economic growth continues to slow. The economics of transport fuels favour the penetration of biofuels over natural gas and electricity; consequently, biofuels grow to account for 10% of fuels in transport by 2060. EVs and petroleum-based hybrid vehicles encompass 10% and 31%, respectively, of the global vehicle fleet. Natural gas increases its share in heavy-duty transport, to 7% of transport fuels in 2060. (World Energy Council 2016, 82)

Here, insufficient investment in infrastructure, low economic growth, and the general tendency to favor biofuels over natural gas and electricity mean that while there is growth in the use of electricity for transportation in the short term, electric vehicles do not receive a similar boost as in the other scenarios in the long term, and thus the rise in use of electricity in transport is less significant.

Unfinished Symphony, Modern Jazz, and Hard Rock—the scenarios that the World Energy Council experts had built and narrated in such a very long-term process—had been reduced to static charts and numbers. Additionally, now that the scenarios had been converted into models, the three plausible narratives had become possibilities. That is, potential paths were converted into actual existing definite and finalized events (from becoming to being, to use Deleuze's terms).

The first task given to the groups was to examine the preliminary results shown in the quantitative models, in search of specific aspects of the narratives that they had worked on up to that point. The participants were also instructed to examine to what extent the variables included in the tables reflected the larger trends presented in the narratives. The participants then tried to find in the models the trends that had been developed in the narrative of the scenarios they had written beforehand. At that point, the models became a kind of reality that guided them as they sought to identify the narratives that they had prepared earlier.

The participants were not sure what to do with the models. After working with and through narratives, with the emphasis being on qualitative thinking and storytelling, they found themselves suddenly asked to translate not only information or events (of the scenarios) but also their entire mode of thinking—which at that point had already been set to embrace plausibility and uncertainty—into a quantitative mode, into a limited set of possibilities with a definite structure. Hence, the participants complained not just about data or information missing in the models but about the entire way of practicing and experiencing the uncertainty of the future. They were confused about how they should think with these models and what exactly their purpose was. While the quantitative models reflected an outcome, an end result of a given scenario, producing a static picture of a particular possibility, the qualitative scenario narrative at the World Energy Council was, as Lorie Taylor put it, "a disruptive framework" that expressed a dynamic process of unfolding events and impacts as well as challenging people's own assumptions: "Scenarios help people understand the assumptions that they are making—that they have a story in their head, but they are blind to it. They take its stakes for granted, and never stop to think about it."

This process, as Nick Mayer noted, has an "element of iteration" that aligns the narratives and the modeling. Accordingly, in extreme, unfamiliar situations where the resulting model is "believable" yet seems impossible in the narrative, the scenario may be modified in accordance with the model. By contrast, Mayer added, while "we probably have reasonably good economic models" for doing short-term work on the energy system in relation to the economy, "nobody knows how to properly filter in the consequences of COVID-19." Therefore, the "interplay" between narratives and models is necessary.

At the last workshop in which I participated during 2018–2019, an artist was also present in the room during the scenario narrative-building process. Throughout the entire workshop, while we were building the narrative of the stories, he would draw on a sheet of paper on the wall, putting into images our language and discussions about the three scenarios, producing an illustration of our conversations (see figure 4.7). Although the artist began drawing his picture from the left side of the paper and worked slowly toward the right, the only chronicity displayed in the picture lay in the order of the workshop itself, as the artist was free to move backward and forward throughout the process, to update and change the drawing as the discussions progressed. Moreover, the image produced by the time the workshop was drawing to a close was not a singular picture of a possible future scenario but rather an illustration of the process of making the scenarios— the language, the images, and the discussions that took place throughout the workshop. Key terms of each scenario could be seen in this picture, whose elements, taken together, illustrated both the complexity of the narratives and the process of producing them.

FIGURE 4.7. Joshua Knowles, conference illustrator, captured the conversation (with ink on paper) in real time throughout the World Energy Council's scenarios workshop, London, 2019. Courtesy of the World Energy Council.

For example, at the top left of the picture, initial premises are presented: "How do we make productive uncertainty?" Related slogans suggest that "[there are] significant changes [and therefore] we need to move not just quickly but differently." At the bottom left, one can see someone in front of a radio, trying to catch signals of change, such as "Trump," "Brexit," "yellow jackets," and "China political shift"—issues that were debated in the workshop in many of the participants' examples. At the center of the picture, in larger figures, some of the key drivers of change are presented, such as "climate activism" and "digital revolution."

The present (the scenario-planning process) and the future (the creation of plausible scenarios) are illustrated through this image simultaneously. This happens, whether by design or otherwise, through a dynamic process in which it is unfinished images, painting practices, and artful images that take part, rather than diagrams, graphs, and numbers. Moreover, in this illustrative model, participants were not external to the technique, as is the case when premade models (created by "external experts") are used. Here, the participants' active role in creating the future narratives affected the design of the model (i.e., the picture).

To sum up, then, the scenario-planning process (at the World Energy Council) worked as an education process, built around the experience and involvement of the participants, in such a way that the result of the process was a change in how participants thought about the future and how they engaged with its uncertainties. As one scenario expert told me about this scenario technique, "It is a recursive tool to create subjectivity." That is, the subject and the (modes of) knowledge-making and power (i.e., veridiction and jurisdiction) are mutually linked. While the participants take part in building the narratives of the scenario through a particular way of thinking about the future and practicing its uncertainties, their mode of subjectivation is also shifting at the same time. What was created in this narrative-based scenario process was a new subjectivity, one that addressed and accepted the potential uncertain future with the idea of multiplicity, heterogeneity, and dynamism as well as by creating a new language and framework for addressing that future.

Reflecting on the nature of his work in an interview published under the title "Questions of Method," Foucault (1991, 75) presents his academic enterprise as an inquiry into "regimes of practices," commenting that "to analyze 'regimes of practices' means to analyze programmes of conduct which have both prescriptive effects regarding what is to be done (effects of 'jurisdiction') and codifying effects regarding what is to be known (effects of 'veridiction')."

Here, it should be noted that in using the term *regime*, Foucault is not alluding to the ways in which new rationalities of governance, as ideologies, are translated into concrete practices. Quite the opposite, he is referring to how different forms of jurisdiction (power) and veridiction (knowledge), along with their as-

sociated practices, affect the emergence of new rationalities—that is, how rationalities are generalized from practices, strategies, and "solid effects" (Foucault
1991, 81).[2]

Such a perspective suggests that if we are to understand uncertainty as a rationality of governing, we need to inquire into its related practices and strategies and to analyze both how they work and whether their forms express a new
rationality (in this case, one that is distinct from that of risk). In addition to
studying the effects of veridiction and jurisdiction, however, we also need to look
at the relationship between them and the effects of subjectivation—that is, how
a new subjectivity is generated from these effects and from the experience of new
modes of veridiction and jurisdiction.

This third element of governing, that of "the subject" (or subjectivation), was
something that became more explicit in Foucault's later work, as he explains in
the preface to the second volume of his *History of Sexuality*: "It appeared that I
now had to undertake a third shift, in order to analyze what is termed 'the subject.' It seemed appropriate to look for the forms and modalities of the relation
to self by which the individual constitutes and recognizes himself *qua* subject. . . .
I felt obliged to study the games of truth in the relationship of self with self and
the forming of oneself as a subject, taking as my domain of reference and field
of investigation what might be called 'the history of desiring man'" (Foucault
1990b, 6).

It was not that Foucault came to see the subject for the first time at this stage
of his intellectual inquiries. Rather, he began to be interested in it as an analytical category. As Kelly (2010, 85) argues, "his previous work did not deny the existence of the subjective dimension, just the analytical need to include it."
Foucault suggested that there are processes of subjectivation in which the subject becomes an object of knowledge. In this regard, Foucault rejected the idea
that the goal of an analytics of government is to show how humans are liberated
or controlled by government. Instead, he argued that government works both
through states of domination and through practices of freedom, which he termed
forms of subjection and *forms of subjectivation* (Dean 1999, 46).[3]

Accordingly, by analyzing the role of the subject in uncertainty-based technologies, and in scenarios in particular, I neither position this work within a discussion of good or bad governance nor address the question of how necessary
the scenario technology is. Rather, I am interested in the mode of experience
expressed by and produced through the technology of scenarios and how actual
practices and strategies (exercised through this technology) promote uncertainty
as the basis of a new mode of subjectivation.

Studies of governmentality that look into subjectivation have mainly discussed it in relation to neoliberal mechanisms (on resilience, for example, see

Brassett and Vaughan-Williams 2015; Chandler and Reid 2016). In this context, subjectivation is presented in scenario-based exercises in terms of the controlled, programmed, powerless subject that is produced through the use of such a technology (Adey and Anderson 2012; Grove 2012; Kaufmann 2016; Reid 2012). Moreover, in these studies, if subjectivity is discussed at all, it is presented only as the *practice* of such technologies. That is, the control exercised over the subject is shown to be a result of his or her use of the technology. In the case examined here, however, the creation of subjectivity is not external to the production of the technology of governance (whereby the subject, or the "social," is understood as coming after the fact, after the technology). Rather, in this chapter I showed how this mode of subjectivation is created in the process of building scenarios and planning for future uncertainties. Hence, subjectivity is shaped from the very outset as part of the process of how the scenario narratives are created and used.

Other studies that have looked at the issue of subjectivation and governmentality have argued that instead of a subject that is subordinated to a disciplinary logic in the sense of "orders" or "sanctions," the production of a subject that is capable of living with (an abstract concept of) uncertainty is also promoted within the governmentality framework (Brassett and Vaughan-Williams 2015; Dean 1999; O'Malley 2010). More specifically, the relationship between the subject, governmentality, and the idea of uncertainty is expressed in the fact that the subject must accept "living with uncertainty" (Pauline Boss, cited in Krasmann and Hentschel 2019, 184).

The current case showed that uncertainty and the commitment to it are expressed in a different way in the scenario technology and its modes of governing. Here, uncertainty is the rationality of a form of action through which the subject thinks and acts in processes of planning for the future. Grounded in this new context, the subject comes to accept the potentiality, flexibility, and openness of the future. In other words, in this technology of governing, uncertainty is not just accepted as an ontological situation (although the uncertainty of the future is accepted as an ontological reality) but also embraced as a modality of practicing and experiencing the world.

SIMULATIONS

Possibilities and Responses

The following newsflash formed part of the WHO's Exercise JADE 2018:

> *News anchor:* Supermarkets have pulled hundreds of fresh tomatoes
> from their shelves as test results provided by the National Food Safety
> Authority have found harmful levels of the bacterium *Listeria mono-*
> *cytogenes* in tomatoes. . . . The National Food Safety Authority strongly
> advises pregnant women, the elderly, and individuals with a weakened
> immune system to not consume the recalled tomatoes. . . . These actions
> come after the declaration of a listeriosis outbreak by the Ministry of
> Health. It is also reported that the number of confirmed cases has risen
> to 100, including 18 deaths. The median age of cases is 61 years. Addi-
> tional deaths are still being investigated to determine if they have been
> caused by listeriosis. (WHO EURO 2019b, Inject 4)

Though this exercise was based on a very specific and detailed story of a food
safety crisis, I would argue that its focus, both in terms of the narrative and in
the actual practice of the exercise, expresses a rationality that is different from
that seen in the cases of the Turning Point and World Energy Council usages
of the scenario technology examined in chapters 2 to 4. Accordingly, I aim to
distinguish between the scenario technology and the closely related technology
of the *simulation*. Though both of these are based on the creation of an imagined
story and its practice in the present, there are important differences between the
two, in terms of the relationship between the fabricated and the real and in how

future uncertainty is observed and related narratives are created, that lead to different roles and different forms of action in each type of exercise.

To date, the growing social sciences literature on scenarios has paid little attention to conceptual and analytical differences between existing mechanisms for governing future uncertainty based on imagined stories, which for the most part are all simply grouped under the umbrella term *scenario*. In this chapter, I seek to rectify this lacuna by highlighting key differences between the scenario technology and the simulation. In my analysis of preparedness exercises conducted by the WHO, I show that scenarios and simulations are distinct forms of governing based on different ways of perceiving the unknown future and different ways of imagining, preparing for, and practicing it in the present. While the scenario is an uncertainty-based technology that embraces the emerging and the unexpected (both in the construction of the narrative and in the actual practice of the scenario), the objective of a simulation is primarily to improve the ability of participants to employ predetermined solutions in the future by addressing a hypothetical, realized event and practicing their response to it.

The failure to distinguish clearly between different types of technologies based on imagined future narratives can perhaps be seen most clearly in the fact that the current literature on "simulations" remains relatively limited (Kaufmann 2016; Opitz 2017). Moreover, when simulations are discussed, the discussion usually focuses on a specific field such as pandemic influenza (Keck 2018; Lakoff 2007; Opitz 2017), disaster management (Revet 2013), or bioterrorism preparedness (Armstrong 2012) rather than examining the simulation technology in relation to or in comparison with other closely related models of future thinking and planning, such as scenarios. In fact, it would seem that the use of the term *simulation* rather than, for example, *scenario* is related more to the type of venue in which the technology is employed than to its particular form of action (i.e., in contradistinction to a different technology of governing, such as the scenario). The term *simulation* is often used to refer to computer-based exercises (Galison 1996; Opitz 2017; Pias 2011; Winsberg 2010), for example, whereas *scenario* is used in relation to field-based, full-scale exercises (Anderson 2010a, 2010b; Armstrong 2012; Schoch-Spana 2004; Simon and de Goede 2015). Thus, the perceived difference between the two types of activity appears to be related more to notions of scale than to the form of power involved, their distinct rationalities as future-governing technologies.

I would, however, like to argue for a more fundamental difference. The scenario technology addresses future uncertainty by using modes of veridiction and jurisdiction that promote the unexpected and reject the rationality of predetermined possibilities that is often expressed in risk-based technologies. However, the premises upon which the simulation technology is based are different, both in terms of how the narrative is constructed and in relation to the form of action and practice of the actual exercise. As we will see, the narrative of simulation exercises relies on a general and universal story that is provided to all participants and contains no specific references that might identify a particular place. In many ways, in the simulation technology, participants remain external both to the process in which the narrative is built and to the specific information provided, the implicit objective being to ensure the separation of the story from the "real" reality—that is to say, the exercise should remain merely a simulation. In terms of the form of action that is seen in the simulation, participants practice predetermined possibilities and work toward increasing the automation of a preplanned response. Accordingly, the scenario and simulation technologies differ quite considerably in terms of how the narrative is developed and how open-ended it is, as well as in the kinds of uncertainty and unexpectedness seen in practice in each case. While the scenario technology addresses future uncertainty by using modes of veridiction and jurisdiction that promote the unexpected, simulation-based exercises are generally designed simply to practice preordained responses and a certain understanding of unexpectedness, particularly in situations where participants have a limited set of possibilities to choose between.

World Health Organization
JADE Exercises

Since 2005 and even more so since 2007, WHO member-states have been legally obligated to implement the standards and capacities required by the International Health Regulations (IHR; see WHO 2005) that were published in 2005 and entered into force in 2007.[1] Compliance with these requirements involves the creation of countrywide capacities for detecting, evaluating, reporting, and responding to potential Public Health Emergencies of International Concern (PHEICs) at all levels of government. Member-states are also legally obligated to report any such incidents occurring within their borders to the WHO. As stipulated in the decisions of the World Health Assembly, the WHO will work with member-states and various international actors, including a range of donors and networks, to assist individual countries to develop their baseline capacities for carrying out such tasks. To enable and organize this work, in 2015 the Strategic Partnership for IHR and Health Security established an online platform known as the Strategic Partnership Portal.[2] This platform aims to monitor member-states' progress with IHR capacity development and to assist them to identify needs, gaps, and priorities at different levels. Among other things, the Strategic Partnership Portal provides materials and tools that have been developed as part of the

WHO's IHR Monitoring and Evaluation Framework, including those used for simulation exercises. Such exercises are supposed to provide an evidence-based evaluation for the monitoring, testing, and capacity-strengthening of countries' ability to respond to disease outbreaks and public health emergencies. Simulation exercises are used to train member-states by exercising emergency response procedures in a safe environment. As a tool for quality assurance, exercises can be used to test and evaluate emergency-related policies, plans, and procedures.

After it was noted that some member-states still had not fully implemented the necessary capacities required in the IHR several years after their entry into force in 2007, the WHO began to develop schemes for monitoring, evaluating, and aiding countries to improve their IHR core capacities, with an emphasis on transparency and the ability to respond promptly. To advance the implementation of the core capacities outlined in the IHR and as a way of testing their function in practice, member-states were also invited to take part in various activities, including simulation exercises designed to prepare participants for and enable them to practice response procedures for potential PHEICs.

As part of this development, the WHO's Western Pacific Regional Office has organized Crystal exercises on an annual basis since 2008 (with the exception of 2009, because of the actual H1N1 pandemic that occurred in that year). These exercises have aimed to strengthen and test the functions of the IHR National Focal Points (NFPs) and the WHO IHR contact points for member-states in the region and have evolved over the more than ten years in which they have now been held.[3]

At the end of each exercise, reflections by participants and organizers are collected so that they might be incorporated in the next exercise. As a result, both the scope of the exercises and the involvement of stakeholders in them have grown over time. In addition, the types of narratives used in the exercises have also changed over this period. Such changes have taken place not only because of comments made by organizers and participants and the need to test different IHR NFP functions but also in response to actual events: an emphasis on information-sharing was seen in 2010 following the 2009 H1N1 pandemic, and an Ebola exercise conducted in 2014 (while an actual Ebola outbreak was ongoing in West Africa) was followed by an emphasis on travel and trade restrictions (an issue that was raised during the Ebola outbreak) in the 2015 exercise that simulated an H7N7 outbreak.

In 2018, the WHO's Country Preparedness and IHR team in the Regional Office for Europe developed a specific simulation exercise, the Joint Assessment and Detection of Events (JADE), for the NFPs of member-states of the WHO European region. Conducted for the first time in November 2018, Exercise JADE was based on the model of the Crystal exercises (conducted by the WHO's Western Pacific Regional Office) designed to test functions, cooperation, and communication between contacts at the WHO regional office and the health ministry representatives at the national level of individual member-states.

Choosing the Scenario: Stimulating Response

Exercise JADE was a "functional simulation exercise." While the exercise was managed from the WHO Regional Office for Europe, participants attended it remotely from their home offices through platforms such as e-mails and phones. The scenario and injects (i.e., short instructions distributed to participants; for example, see text box 5.1) for the exercise were based on five objectives: to test communication between the NFPs of WHO member-states and their regional contacts in the WHO, to test NFPs' access to and use of the Event Information Site (EIS), to practice and test NFPs' assessment of public health events using Annex 2 of the IHR and its notification process (including posting on EIS), to assess multisectoral coordination between NFPs and other relevant sectors in the conducting of basic risk assessment, and to assess coordination between NFPs and communications personnel in the development of statements and communication material for the public.[4]

Accordingly, at a meeting to decide on the scenario that would be used in the exercise, discussions focused less on any particular questions of content than on whether the scenario would serve the above objectives. The scenario expert who attended the meeting connected his laptop to the large screen and presented several options (e.g., zoonosis, cat flu, and an earthquake). One of the organizers, looking at the different options, suggested a food safety scenario could be suitable since "we have a person who is an expert on that." The scenario expert then played a video, a food safety scenario inject. The organizers agreed this scenario could be useful, and one of them noted, "Listeria is a burning issue so it's relevant. . . . We even have a multi-country event open." After several questions were asked to ensure the scenario would fit with the objectives of the exercise, it was decided that this scenario would be used.

As indicated by the speed with which it was selected in the above process, the actual story in the exercise was not particularly important in itself but was viewed rather as a stimulant for certain responses that would serve to achieve the exercise's objectives. In this and other simulation exercises, then, participants are occasionally reminded that the narrative being presented is hypothetical so that they stay engaged less with the story itself and instead remain focused on the simulation's practices and on following instructions toward particular responses:

> Please recognize that this exercise will use a simulated, artificial scenario that *may not reflect a real-world situation.* Players should accept these artificialities. Please do not be overly concerned by complexities or details associated with the exercise scenario itself. The objective is to work with the scenario *to facilitate your actions for communications, rather than to challenge it or seek to resolve every last possible detail.* (WHO Western Pacific Regional Office 2011, 31, emphasis added)

Thus, while the narratives in the scenario technology discussed earlier (in Turning Point exercises or in the World Energy Council's scenario-building) considered almost all possible information about the present and the future to make the stories as realistic as possible, the stories in these WHO simulations do not reflect "a real-world situation." What matters in the case of the WHO simulations is not the content of the story or knowledge of the way that story unfolds but rather the responses of the participants to the story, with the latter being encouraged to "facilitate . . . actions for communications . . . rather than . . . to resolve every last possible detail." In other words, the story is designed to help the participants engage but also directs them away from the details of the narrative and toward the goal of communication and response rather than toward understanding the root cause of the event itself or related emerging problems.

A few months before the start of Exercise JADE, all participants received an e-mail invitation with information on the purpose, scope, and date of the planned exercise. Next, a week before the exercise began, a first inject (Inject 0) was sent out to participants (together with the Participant Guide) (WHO EURO 2019a, 4). This first inject took the form of an e-mail sent from the exercise management to the NFPs, and provided details of social media and local press reports regarding the "event." The e-mail mentioned reports of an unidentified GI (gastrointestinal) infection affecting residents at a nursing home (some of whom had been hospitalized for reasons such as dehydration) and the nearby area. The e-mail also reported that affected individuals were experiencing confusion, hallucinations, nausea, vomiting, diarrhea, and flu-like symptoms. Participants were informed that three deaths had been reported in the media, but it was pointed out that these had not yet been confirmed. The cause of illness was described as being as yet unknown. On receiving this inject, participants were asked to confirm that they had received the e-mail. From then on, the exercise gradually evolved from inject to inject, with each inject providing specific information and a request for specific action.

At the start of the actual exercise, Inject 1, dated 1 June 2018, was sent out to all participants. This was a copy of the inject that had been sent a few weeks earlier (see text box 5.1). It included an e-mail from the exercise control, notifying the participants (NFPs) about media reports in their countries and providing examples from different social media sites and a local newspaper. The background story for the inject was as follows: "Reports spread on social media (Inject 1) of an unidentified GI infection affecting the residents of a nursing home called Sunflower Nursing Home. Using the hashtags *#SunflowerDisease* or *#SunflowerNursing,* people on social media report symptoms such as confusion, hallucinations, nausea, vomiting and diarrhea, as well as flu-like symptoms. Three people have died according to social media reports, none of the deaths have been confirmed. A news article in *Truth News* is published summarizing the events" (WHO EURO 2019a, 20).

Text box 5.1. Email from exercise control, Exercise JADE Inject 1. Source: WHO EURO 2019.

SIMULATION   SIMULATION   SIMULATION
This is in YOUR COUNTRY

From: Exercise Control (EUROcontrol@who.int)
To: Your Country's National Focal Point
CC: EUROcontrol@who.int
Date: Start of Exercise (1st June 2018)
Subject: Reports from local press and social media. Possible alert

THIS IS AN EXERCISE MESSAGE
Background information on scenario

Social media and the local press in your country reported on a number of signals related to cases of confusion and gastrointestinal (GI) infection amongst residents of a nursing home. Users of social media are posting comments and messages about their parents and grandparents being hospitalized with GI symptoms and confusion due to an unknown disease. Most of the status updates and comments posted show what appears to be a type of GI infection causing dehydration, confusion or sepsis in older patients. The cause of these symptoms is currently unknown, but many relatives blame the poor quality of care at the nursing home.

Thus far, it is unclear how many are affected, but many of the cases referred to on social media seem to be residents at the Sunflower Nursing Home in Nearby Province (the province 122 km from your capital). The people posting appear to live in relatively close proximity to each other, with a few exceptions, or have tagged Sunflower Nursing Home in their social media posting.

Reported symptoms include confusion, weakness, fatigue, nausea, muscle aches, diarrhea and vomiting. Some residents have been hospitalized, seemingly due to dehydration, altered mental status or sepsis. The pathogen has not yet been identified. Three deaths have been reported through social media using the hashtag #SunflowerNursingHome—none of these have been confirmed.

——END OF INJECT—

ACTIONS TO TAKE NOW
Please acknowledge the receipt of this email and the start of the exercise by replying to <u>EUROcontrol@who.int</u>

The information about the possible emergence of a new outbreak was not followed by instructions for medical action or to conduct an outbreak investigation. Rather, the only type of engagement required from the participants was that of confirming that they had received the e-mail itself. Participants were not asked to identify or diagnose what was happening or to describe an appropriate medical response. Instead, they were asked to respond to the information in their capacity as NFPs, technically, using the appropriate forms of communication between the reporting bodies and the recipients.

Narrative as Fabrication: The Relation to Reality

As noted above, Exercise JADE was first conducted in November 2018. Over two separate days (13 and 15 November), all of the countries that had agreed to participate in the exercise were tested over a five-hour period. On the first day, only English-speaking participants took part in the exercise; on the second day, only Russian-speaking participants were involved. A total of twenty-six countries participated in the exercise over the two days, with each country following the exercise from its local NFP office.

The WHO "management team" for the exercise consisted of eleven people who together organized, executed, and evaluated the exercise. This team included staff from the WHO Health Emergencies Programme at the WHO's Regional Office for Europe and their colleagues from the WHO Health Emergencies Programme Country Preparedness and IHR Unit at the WHO's international headquarters. Besides providing support for the conduct of exercises such as Exercise JADE, the Health Emergencies Programme Country Preparedness and IHR Unit at WHO headquarters also develops exercise documentation (including injects, handbooks, and action sheets). The specific scenario narrative used for JADE (initially intended for an exercise in New Guinea) was developed at the WHO headquarters by scenario experts together with WHO food safety experts.

The narrative for Exercise JADE was based in a place called "Your Country," which was used to indicate each of the countries participating in the exercise without specifically identifying any particular country. Another name used in the exercise was "GlobalLand." This fabricated name was given to a country wherein additional, related events took place outside the country of the participant (i.e., a neighboring country) yet elicited a response from them (as discussed above). The date given for the start of the event was 1 June 2018 (which then lay some five months in the past, as the exercise itself was held in November 2018). At that point in time, according to the simulation narrative, reports of a mysterious gastrointestinal infection (GI) affecting residents at a nursing home were beginning to circulate in social media and the local press: "Parents and grandparents [are] being hospitalized with symptoms of GI and confusion due to unknown disease. . . . The cause of these symptoms is currently unknown" (WHO EURO 2019b, Inject 1).

Two days later, according to the exercise time frame, a newspaper article suggested a possible connection between the spread of the illness and a mass-gathering event (an international festival) that many of the nursing-home residents had previously attended (WHO EURO 2019b, Inject 2). Several days then passed, and on Day 7 the Ministry of Health (of "Your Country") published a first press release on the issue, declaring that the mysterious infection was *Listeria monocytogenes*, and that eighty-eight cases and sixteen deaths had been confirmed. The source of the outbreak, however, was still unknown at that point in time. An e-mail from the Ministry of Tourism and Economy to the Ministry of Health expressed concern about potential risks to tourism and trade and asked whether there had been reports of any similar events in other countries (WHO EURO 2019b, Inject 3).

This story, it should be noted, was sent out, inject by inject, to all of the countries participating in the exercise. Accordingly, while detailed, it was kept sufficiently generic and universal to be relevant to all of them. It was divided across thirteen injects, which were sent out gradually as the exercise progressed and together comprised the pieces of the puzzle that would make up the whole event. The entire story was developed throughout the five hours of the simulation, setting out what happened during the first thirty days following a disease outbreak in "Your Country," and, in the final inject, skipping to the aftermath of the outbreak several months later.

As we have seen in chapters 2 and 3, in the case of narrative-building for the Turning Point scenarios, concrete and specific information related to the different units that would be participating was taken into account in the design of the exercises. The narratives for the Turning Point scenarios were thus initially based on realistic analysis and accurate information, provided by Israeli intelligence agencies, which referred to present and past events in Israel. However, the scenario narrative for each new exercise would also break with this known information to create something new, a new re-mediated imaginary story. This new reality, this imagined yet-to-be-real event, was an important characteristic of the scenario, since it contributed to the real experience of the participants in the exercise, as well as to keeping them involved and challenged each year, thus encouraging them to act in the exercise as though it were a real situation.

In the case of JADE exercises, however, the narrative referred to "Your Country," a hypothetical place, and contained no concrete references to any particular place in the real world. The initial description that set out the background for the simulation was followed by information provided through injects (documents and movies) that were also very general in nature (though they might be related to a range of different types of events, such as zoonosis, flu, earthquakes, etc.). In principle, then, the accuracy and realism of the story used in the exercise had less significance than the actions it was supposed to prompt. The importance of this was also expressed in the context of some of the Crystal exercises, where evaluators' feedback regarding the exercise included remarks such as the following:

> Some exercise participants did not use "Exercise Exercise Exercise" or "This is an exercise" in each message [as required].

There was some confusion with the telephone call made to test communication. Some participants thought that it was part of the scenario.

Some participants did not understand when the exercise ended or were confused that the scenario was real.

Some participants were fighting the scenario. (WHO Western Pacific Regional Office 2011, 21)

Similarly, in the WHO's JADE exercise, every inject sent to the participants was marked as being part of a "simulation" in its title, to emphasize the fact that the information related only to an exercise and not to a real event. Unlike Turning Point and World Energy Council scenarios, which use the names of real newspapers and journalists when presenting media news to make the exercise narrative as realistic as possible, in the WHO simulation names are fabricated—for example, "GlobalLand," "*Truth News*," and "*News First*"—which makes it possible both to use the same scenario narrative in a generic manner (across countries) and to remind participants that the situation with which they are engaging is only an exercise (as well as to prevent possible misunderstandings in a case where someone who is not participating in or aware of the exercise accidentally receives an inject).

The general nature in which the narrative for the simulation was presented, including the generic details provided about the affected country, is linked to the relationship between fabrication and realism in the simulation technology. While, as we have seen earlier, the goal of the scenario technology is to create a narrative that is not only reasonable but also very *realistic*, enmeshing reality and fiction so much so that the behavior of participants is at times marked by the tensions and stresses that they might experience in a real event, in the current case the design of the simulation narrative (while certainly intended to have an actual effect in the real world) is supposed to help participants keep in mind the *distinction* between the real reality and the "fictional" world in which the exercise and actions are taking place.

Exercising: Practicing Predetermined Solutions

Although the simulation's narrative was unfolded through the provision of specific details regarding how the disease outbreak began and might potentially spread throughout the world, in terms of practice, the activities of the participants in relation to this narrative were oriented toward engagement with potentially key personnel from their own countries (e.g., Ministry of Agriculture staff) or from international bodies (e.g., International Food Safety Authorities Network members) (whether they actually joined the exercise or were simulated by someone from the WHO) rather than toward understanding and identifying the event itself or solving the actual health crisis. As a WHO technical officer commented in a discussion about the scope of the exercise, "The minimum is just knowing who to contact. . . . More than that would be to ask communication people to participate."

In this regard, every inject or a piece of information delivered to the participants ended with very specific instructions about the required response. For example, in Inject 2b, a simulated journalist e-mailed the participants, asking them to provide the following information: "Briefly outline the next steps that the Ministry of Health will take in communicating with the public, i.e., will a press statement be made. Are you aware of any Emergency Risk Communication (ERC) plan and SOPs [Standard Operating Procedures]? If yes, please outline the next steps the Ministry of Health will take according to such a plan. If an ERC plan and SOPs are not known or available, outline specific actions that you would take as the IHR NFP, if any."

Participants thus needed to know whether their countries had Emergency Risk Communication plans and/or Standard Operating Procedures in place (and whether they had access to them). Where this was not the case, they should nevertheless be able to provide information on what the next steps would be or at least consider why they were unable to do so. In other words, exercise participants practiced not just a response but also a particular set of possibilities that they should use in their response, that is, ERC plans or SOPs.

Furthermore, all of the information concerning the simulated event was revealed to the participants during the simulation. This was essential for the main goal of the exercise, which was to practice not the content of the specific event itself but a set of standard responses (set out in the IHR guidelines) for any future outbreak. This stands in contrast to a scenario exercise, in which participants have either to create the narrative from the start (World Energy Council) or to reveal the "hidden" story while practicing it (Turning Point).

It was this designed exercise that would make it possible for participants to practice a predefined set of actions. As one individual involved in preparing the exercise summarized the process, "So Inject no. 1 is background information—the objective is that the exercise started and that they received the material. . . . The next inject would be what they respond to. Just a starting background. . . . Now we can show them the video and ask for verification. Ask them to assess and whether they have additional information. They will have to say if it's according to IHR criteria and explain why. . . . You should give them enough information to make a decision. Request national risk assessment using the template provided [in Annex 2 of the IHR]."

The inject for the third day of the simulated story (Inject 2) included various items: a newspaper article published by the fictional *News First* together with summarized special media reports. The newspaper article mentioned that many of the affected residents had attended a large international festival (the Mozart Boogie Woogie Festival) and that no official statements had been made by the health authorities so far.

The inject also included an e-mail from the *News First* journalist asking for official information about the reported cases. In this e-mail, the journalist first asked for contact details for the press office person at the (local) Ministry of Health and then asked how the NFP representative decides whether an event is of interest to the WHO. In response, exercise participants were required to deliver contact details for the relevant government personnel with responsibility or authorization to provide official answers to the media. They then needed to specify the conditions under which they, as NFPs, must report a given event to the WHO. The first request thus repeated the goal of the first inject regarding communication of information and testing. The second seemingly moved toward the examination of decision-making processes. However, although this question sounds open and hypothetically could be answered differently by different participants (according to their different understandings of the situation or different levels of responsibility), in fact the required response is subject to the understanding of individual participants only insofar as it is founded upon and addressed through the IHR instructions. That is, the participants were directed to respond by exercising a very specific form of decision making that is set out in Annex 2 of the IHR (as presented in the schematic in figure 5.1).

Similarly, the set of actions requested were also provided in advance. For example, Inject 4a read, "Please assess the current situation by using the IHR (2005) Annex 2 decision instrument and advise if the current situation warrants notification to WHO. If yes, please do notify WHO and please let me know the conclusion of your assessment and actions. . . . Please inform us which stakeholders should be involved in the decision-making, and kindly provide us with their updated contact details."

The first request shows once more that the entire exercise is about practicing a standardized decision-making procedure rather than being oriented toward the content of the event and its particular nuances and future unfolding. Moreover, it does not matter how detailed the story is or how it develops in terms of the health/medical content; what is practiced in the simulation is the appropriate response, which is understood in terms of participants being able to follow

ANNEX 2

DECISION INSTRUMENT FOR THE ASSESSMENT AND NOTIFICATION
OF EVENTS THAT MAY CONSTITUTE A PUBLIC HEALTH EMERGENCY
OF INTERNATIONAL CONCERN

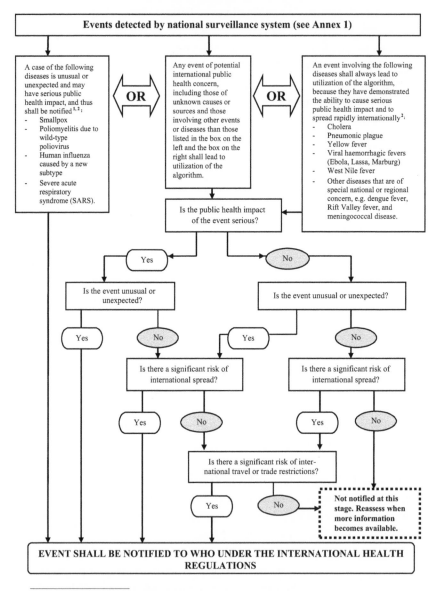

**EVENT SHALL BE NOTIFIED TO WHO UNDER THE INTERNATIONAL HEALTH
REGULATIONS**

[1] As per WHO case definitions.
[2] The disease list shall be used only for the purposes of these Regulations.

FIGURE 5.1. Decision instrument in Annex 2 of the *International Health Regulations*. Used by permission of the World Health Organization (2005, 43).

Text box 5.2. Email from Ministry of Health, Exercise JADE Inject 3. Source: WHO EURO 2019.

SIMULATION   SIMULATION   SIMULATION
This is in YOUR COUNTRY

THIS IS AN EXERCISE MESSAGE

Dear colleagues at the National IHR Focal Point,

We have now received confirmation that the recent outbreak in Nearby Province is caused by *Listeria Monocytogenes*. Please see lab results at the end of this email. So far, 88 cases have been confirmed, and 16 people have died (all of these were at-risk citizens with other comorbidities).

We have received an inquiry from the Ministry of Tourism and Economy who are concerned about how the current outbreak may affect tourism in the country. They have asked us whether any other listeriosis events have been reported from other countries.

Please take the following actions:

- **Please provide the MoH (EUROsimulatorB@who.int) with a list of the first 3 current events on the Event Information Site (EIS)**
- **Please verify if your NFP contact details are correct and share those**
- **Based on EIS, provide information on global listeriosis outbreaks since January 2017 (total number and countries those were created for)**

Kind regards,
Dr. Salem O'Nella
Director, Public Health Unit
Ministry of Health

the rules, the procedure that has been created in advance, and to act according to prewritten manuals and guidelines (see, for example, text box 5.2).

According to the schematic in Annex 2 of the IHR, NFPs should report events occurring in their country to the WHO when two out of four of the following criteria apply: there is a potential serious public health impact, the event is unusual or unexpected, there is a significant risk of international spread, or there is a significant risk of international travel or trade restrictions.[5]

Hence, we see that the simulation is directed not merely toward exercising responses but also toward a particular set of responses intended to establish patterns of automatic behavior among the participants. This idea of automatization was already present in the Crystal exercise, which was the format upon which Exercise JADE was built: "Crystal exercises have an underlying intention to train and practice specific response behaviors that need to be *essentially automatic* when dealing with public health threats. In this sense, they are not unlike a fire drill in the specificity of expected actions. Consequently, all Crystal exercises have had a similar design and provide similar experiences, with variations in exercise scenarios to capture interest and provide different practice opportunities" (WHO Western Pacific Regional Office 2015, 9).

After conducting an assessment, all participants were asked to "share their initial risk assessment under Annex 2 of the IHR (2005)." That is, they were asked to activate the risk assessment format and to reach a conclusion as to whether they should report to the WHO about the situation at hand. Not surprisingly, of the twenty-one countries that provided a response to this inject, twenty representatives concluded that there was a need to notify the WHO about the situation in their countries. Only one country decided that the event did not require notification on the grounds that it did not fulfill at least two of the four specified criteria. Ultimately, this constituted a problem. The information provided in the injects was supposed to ensure that participants would view the event as notifiable to the WHO according to the criteria set out in Annex 2 of the IHR.

Put differently, the aim of these simulation exercises was to practice responses and test participants' compliance with the required instructions—to the extent that the exercise was designed to avoid any deviance where there was any possibility that a participant might conclude that a different response was required (i.e., a response that differed from that predefined in the exercise objectives). The generic tool provided ensured not only that participants would follow the procedure of how, what, and when but also that their conclusions would be more or less similar (the goal here being to approach "objectivity" through intersubjective expert consensus). Hence, it left little room for interpretations or any unintended developments during the simulation and the practice of the event. Therefore, if for some reason participants went off-script, the exercise team was responsible for bringing them back into line, as the 2015 Exercise Crystal guide for simulators and controllers explains: "Please remember that the exercise will run for five hours (equivalent to three days in the scenario). Do not fight the scenario and, similarly, encourage the players. To facilitate exercise communications, as long as you remain consistent within the provided scenario, please feel free to be creative within the limits of plausibility when providing information requested by NFPs" (WHO Western Pacific Regional Office 2015, 38).

Hence, those facilitating the exercise had to operate within relatively strict boundaries, and could create or develop the exercise only according to the limits of the preplanned narrative and its set of possibilities. In this regard, the exercise controller sought to maintain control over and limit the interpretations and responses of the participants in the exercise:

> The role of the exercise controller is to ensure that the exercise runs smoothly and achieves its stated purpose and objectives. The exercise controller must monitor and adjust the exercise to ensure that the exercise achieves its goals. If the exercise evolves beyond its intended scope, the exercise controller should intervene, either by adjusting *the timing of the injects*, or by delivering *ad hoc injects*, which are additional pieces of information that are developed during the exercise to advance, slow, or redirect exercise play. Participants often make *unanticipated decisions*, and the exercise controller must be able to respond to these decisions in line with the exercise objectives. He or she can be assisted by role players and/or evaluators, who will be assigned to specific injects of the scenario, and/or who will control the flow of participants' interventions. (WHO, n.d., emphasis added)

Moreover, to standardize the exercise dynamics, not only were the content of the event and the possible responses predetermined, but the time period within which participants should respond was also controlled. As the exercise progressed, later injects contained explicit instructions to respond within a given time limit (see text box 5.3). Injects ended with remarks such as "We would appreciate receiving your feedback within 30 minutes. Thank you very much."

The exercise was limited to a period of five hours, during which time all participants should practice all of the planned stages of the exercise. Accordingly, if participants were delayed in their responses to particular developments, they would be requested to move faster in subsequent stages.

While in the Turning Point scenario-based exercises it was up to the participants to create the puzzle of the event (an integrated story), to identify the event's details and to compile the multiple incidents into a comprehensive story, here everything was provided by the WHO simulation exercise management team. That

Text box 5.3 Email from the IHR contact point for the WHO Regional Office for Europe, Exercise JADE Inject 4c. Source: WHO EURO 2019

SIMULATION   SIMULATION   SIMULATION
This is in YOUR COUNTRY
From: HIM, euroihr@who.int
To: IHR National Focal Point
Subject: Accuracy check of EIS post
Data: 26th June 2018

THIS IS AN EXERCISE MESSAGE
Dear National IHR Focal Point,

Based on the current information, this listeriosis outbreak has the potential for international spread and may affect international trade of the contaminated food products. WHO would like to post this event on the Event Information Site (EIS) to alert other National IHR Focal Points in order to carry out appropriate control measures.

Please find attached a draft EIS text for your accuracy check. We would appreciate receiving your feedback within 30 minutes.

Thank you very much.

---

is to say, while in Turning Point exercises participants were only provided with details of incidents that had occurred and then had to collect that information and construct the overall picture of the scenario event, here the diagnosis of the event was fully provided by the exercise management team. Step by step, inject by inject, the full narrative was given to participants, who had no influence over how the events develop.

Unlike the management personnel engaged in the WHO simulation exercises, participants in the Turning Point scenario exercises attempt to understand and diagnose the events that are taking place in the scenario and then to extract from those exercises new possible problems that would not have been revealed if the scenario had not been conducted. Scenario exercises, then, are aimed at problem extraction, while simulation exercises are directed toward testing the concrete and specific aspects of the prewritten event and the realization of its predetermined solutions.

Planning for Future Events: Reassurance and Certainty

In Exercise JADE, on Day 7 of the story (7 June 2018), the third inject provided information about a first press release by the Ministry of Health that confirmed *Listeria monocytogenes*. The report mentioned eighty-eight confirmed cases of the infection and sixteen deaths. It also revealed that epidemiological and laboratory investigations were being conducted to determine the source of the outbreak. In addition, it mentioned an e-mail received from the Ministry of Tourism and Economy that prompted a need to check the EIS platform (to which NFPs have authorized access) to see whether there had been similar outbreaks in other countries. Here, again, regardless of the health information provided, the instructions for action on the part of the participants were that they should access a particular website and report according to the information provided there (the goal being that NFPs should become aware that they needed to access the EIS, that they were technically able to do so, and that they knew how to use it once they accessed it).

By Day 26 of the story (26 June 2018), Inject 4 provided further details of the investigation by the National Food Safety Authorities, which had been sent to the Ministry of Health. Participants were informed that the investigators had found *Listeria monocytogenes* in tomatoes at the New Horizon Farm in Province South, and the strain was identical to the type of *Listeria* that had caused the current confirmed cases. Thus, the New Horizon Farm was identified as the source of the outbreak.

The Ministry of Health issued an updated press release on the outbreak (Inject 4a), confirming the source of the outbreak as tomatoes from New Horizon Farm. Further information and instructions for subsequent actions required from the participants were then provided. In this sense, as the exercise linearly progressed toward conclusion, the full narrative was revealed to the participants and uncertainty was removed.

Evidently, we can see that the idea of uncertainty and indeterminacy built into the scenario mechanism contrasts sharply with that built into the simulation exercise. In the scenario, the uncertain future is developed into plausible future stories that aim to present a reasonable and convincing narrative, so that participants might think through the unknown future and practice their imagination in narrating and exercising—and hence experiencing—unpredictable situations and open-endedness in the structure of the technology.

The simulation technology presents a more predetermined and closed form of planning and exercising toward future events. Instead of uncertainty, which is merely apparent here as part of the attempt to mimic reality, simulations mainly exercise sets of ordered reactions to a particular experience using existing knowledge, possibilities, and structures. In simulation exercises, the purpose is not to narrate or practice plausible future events but rather to use the specified event that forms the subject of the exercise narrative and its unfolding to test specific actions and responses. In other words, a simulation exercise is predetermined and is more about reassurance (Krasmann 2015) and certainty (Tellmann 2009) than uncertainty.

SCENARIOS, TEMPORALITY, AND UNCERTAINTY

The scenario technology involves two separate temporal dimensions, neither of which is linear. The first temporality is "future-now thinking" or the "present future." That is to say, scenarios create future presents by bringing imagined future narratives into the present and enacting them to better extract new problems that would not otherwise have been identified. The second dimension of the scenario temporality is that of the "future present." This refers to the "real" present that will take place only in the future. By acknowledging the imagined future and realizing it in the present and by addressing the future present and accepting its potentiality and openness, the temporality of scenarios works at one and the same time on both the actual and the virtual, while moving away from any linear conception of time. Scenarios create futures as presents but also increase uncertainty, becoming, and potentiality as part of the modality and the technology's way of action. Here, then, I will examine the role of time and temporality within the scenario technology.

Having completed my analysis of three distinct aspects of the scenario technology—its knowledge making, its governing mechanisms, and its subjectivity—I will examine the technology and its temporal dimensions in a more comparative fashion across the different sites or fields studied. The primary focus of my argument will be to show how thinking, imagining, and practicing the future through scenarios is manifested in different ways in the context of security, health, and energy, depending on the particular temporality and the conceptualization of future uncertainty in each field. In addition, I will show how the specific temporality of each site is not only employed in the scenario narratives and exercises but also affects the development of different forms of scenario technologies, each with its own specific orientation toward uncertainty.

Time and Temporality

Moving from the notion of time to that of temporality, scholars have recommended that we differentiate between time as an "objective" figure (past/present/future tenses) and temporality as time perception based on subjective measurements, as well as between historical and social time (Hodges 2008). According to Alfred Gell, "A-series time [is] the subjective, tensed existence involving past, present and future relations that comprises everyday human time perception. B-series time provides the basis for A-series perception. . . . The 'real' world does not exist according to A-series laws of perception. . . . Instead, we know B-series time through temporal models, which reflect the structure of B-series time without accessing it directly" (cited in Hodges 2008, 403–404).

In the following analysis, then, I will highlight different modes presented in the three anthropological cases and how the scenario technology is affected both by time and by the temporality of the field in which it is employed.

Laura Bear (2016) has argued that an examination of the different aspects of temporality (the techne [techniques], episteme [knowledge], and phronesis [ethics]) should form a part of any sociocultural inquiry, as tracing each of these can help deepen our understanding of the problems that we investigate. She also recommends that we examine temporality by looking at different dimensions—such as knowledge, technology, and ethics—that are better observed together and analyzed simultaneously. Likewise, several other studies (e.g., Ahmann 2018; Antonello and Carey 2017; D'Angelo and Pijpers 2018) have put forward the idea that considering different aspects of temporality can serve as a lens through which to better understand society, culture, or the economy.

However, rather than regarding conceptions of time or temporality primarily as a lens through which to inquire into society or to provide methods for preparing for the uncertain future, in this chapter I am interested in a different aspect of the relationship between temporality and technology and particularly how temporality affects the development of social technologies. I therefore first illustrate how each of the three cases examined in chapters 2 to 5 expresses a different temporality and involves a different type of future that is being imagined and prepared for. In addition, I show how the scenario technology is developed and expressed in different ways in each case in response to the specific temporality of the particular field in which it is employed. Finally, I conclude that the relationship between temporality and technology seen here is a mutual one: not only does the technology affect the way in which the future is perceived and governed, but also the time orientations, or temporality models, that are specific to a particular field or site affect the development of the scenario-planning technology in each case. Thus, rather than taking the scenario technology as given or as a set form predisposed to the future, I show how it is developed and used in different contexts or fields, and how it is affected by the specific temporal dimensions of each case.

In addressing the relationship between temporality and the scenario-planning technology, I discuss each case in terms of three analytical temporal dimensions: chronological time, the time of the event, and the temporality of the intervention (as will be explained below). Taken together, these three dimensions influence the development and usage of a particular format of the scenario-planning technique in each of the fields examined, expressing different orientations toward uncertainty in each case.

By *chronological time*, I refer to the positioning of the particular field on the timescale from the past, through the present, to the future. This is an external temporal context to the technology and is directly related to the temporality of each field (security, health, and energy)—that is, the extent to which the future in a particular field is considered as being an immediate present, a recent-past crisis, or a near future. By *time of the event*, I refer to the types of future events the organization is preparing for and how far into the future they are—that is, the temporality of the events created in the scenario, which is the "internal" temporality of the scenario narrative. By *temporality of the intervention*, I refer to the timing of the intervention, whether it relates to the immediate short term or to the long term, and to what extent the process of building and practicing the narrative is directed toward an intervention (i.e., response to these events) in the near or far future. The interactions of these dimensions produce a temporality model that is related to a particular scenario-planning format. Each such format reflects a specific configuration of narrative building, form of practicing, and orientation toward uncertainty. That is, if scenarios express three modes of uncertainty—knowledge, power, and subjectivity—this configuration is also affected by the particular temporality within which it is embedded and created in each field.

The contemporary landscape of global public health is predominantly characterized by a form of crisis-oriented perception (see Lakoff 2017), in which the future is situated in relation to the already known past and the currently occurring present. Scenario planning activities in this field are thus developed to deal with urgent actual present events or to prepare for these types of events by improving capacity to respond immediately with a set of known practices. Accordingly, the mode of intervention in the WHO is oriented toward practicing that which is already happening and reacting swiftly to events that are taking place. That is, participants do not take part in a long process of building the narrative for the scenario and instead practice (preexisting) solutions that have been shaped and driven by known past events. Rather than an open format aimed at extracting new problems (as in Turning Point exercises or the World Energy Council's scenario-planning activities) and increasing the unexpected and the unknown, here the scenario technology functions as a simulation that generates the experience of an immediate present crisis and, only as part of this, a structured sense of a slowly diminishing uncertainty.

Within the energy sector, scenarios are used to imagine and create long-term energy future stories. Additionally, scenario planning is deployed toward potential futures, which are related to the past and the present, but seeks to move away from the known toward the unknown—and even the unthinkable. This orientation also affects the type of intervention and the form that the scenario planning activities take. Here, emphasis is placed on the importance of the long process of narrative-building as well as on the need to involve participants in the creation and writing of these future stories. In this process, the narrative building becomes also a mode of intervention, since it shifts not only the kind of future stories created or the actual practices and responses to them but also the subjects' mode of observing and accepting the uncertain future. Subjects imagine themselves in the future and then create new events.

These events are open to dynamism, indeterminism, and change. Such a dynamic process not only makes it possible to detect new future problems but, through the thinking and imagining of multiple future events, also affects the understanding of the future as uncertain and nonactualized potentialities. Put differently, in the World Energy Council's scenario-planning workshops, the scenario narrative format is designed yet never finalized; it is always "in becoming," actualizing into particular stories but always with the potential to shift again according to current and future uncertainties.

Turning Point exercises emerged as a response to the problem of emergency preparedness in the context of border security threats. While the time of the WHO's health scenario exercises is oriented to the urgent and immediate present, and that of the World Energy Council's energy scenarios is set to the long-term future, these security scenario exercises are directed toward the near future. The narrative-building process is based both on past information and on new events that are re-mediated, bringing together the old and the new into a shared future story. Moreover, narrating and practicing are two essential parts of this scenario technology format. That is, the exercises involve both the creation of plausible scenarios and their subsequent practice in the present, in a process that enables the identification of new future problems.

Health: Experiencing Past Events through Structured Uncertainty

Exercise JADE (WHO EURO 2019a) was developed as part of the WHO's efforts to assist countries in implementing the IHR (and, as explained in chapter 5, the use of exercises and other activities for implementing these regulations relates to the Monitoring and Evaluation Framework). Generally, such exercises are used as a way to monitor, test, evaluate, and strengthen countries' capacity and ability to respond to disease outbreaks and public health emergencies. Like various other simulation exercises carried out by the WHO, JADE has been used to train relevant personnel in the organization's member-states through the exercising of emergency response procedures in a "safe" environment. Designed on the basis of the Western Pacific Regional Office of the WHO's annual Exercise Crystal, JADE also makes it possible to test and evaluate emergency-related policies, plans, and procedures.

As I showed in chapter 5, simulation exercises like JADE are directed toward practicing a particular set of preordained responses and allow very little room for sudden changes and developments in the script. The idea is that as a result of practicing preordained responses, participants will be better able to perform their roles rapidly and appropriately in an actual future event. As part of this process, for instance, participants check that they have no technical problems (e.g., by making sure that their e-mail communications are functioning properly), ensure they have clear guidance regarding whom they should contact in an actual event (e.g., the regional WHO contact point) and how, and practice conducting a risk assessment.

Simulation exercises in this context are built around the idea of the occurrence of an immediate present threat. Thus, in these exercises, scenarios present situations that are set in an immediate-present urgent temporality. This immediate-present perspective is maintained throughout the entire exercise, even as each new inject moves participants into the next stage of the story line, informing them of how much (scenario) time has passed since the last inject was received (sometimes a day, sometimes entire weeks or more). In this sense, simulation scenarios are not only an instrument for bringing possible uncertain futures into the present to practice known solutions for them (as noted in chapter 5) but also a way to experience the immediacy and urgency of a crisis, with its uncertain possibilities, as it unfolds in the present. Put differently, here the time of the event is set in a way that is intended to generate an experience of an immediate present that could unfold in different, uncertain ways and must therefore be acted on urgently.

In the implementation of the exercise itself, to create such an experience and to get participants to act as though an actual event requiring an intervention is taking place, the simulated present reality (the crisis in these simulation exercises) is designed to mimic the conditions of the actual present reality. Indeed, even the technical means of communication used in exercises might need to be updated to enable this—as, for example, when fax machines were retired from use and e-mails instead became a central communications tool (a change that needed to be made in Exercise Crystal, which has been running for over a decade; see World Health Organization Western Pacific Regional Office 2011, 2015). Nevertheless, it is important to note that a more accurate claim would be that simulation exercises are purposely designed to nearly mimic actual present reality, thus maintaining a clear distinction between two spaces of intervention: actual present reality and simulated present reality.

In Exercise JADE, the scenario was selected from a set of ready-made scenarios prepared in advance by a team of experts at the WHO headquarters (and then modified to fit the specific context of the exercise). Such ready-made scenarios are usually focused on a specific kind of hazard (e.g., a cholera outbreak, an earthquake, a nuclear accident, or a civil conflict) and are supposed to be sufficiently realistic to provoke the type of response that would be required in an actual event. This is achieved, for example, by using prerecorded news media videos reporting the event. Similarly, in terms of content, the scenarios are realistic in the sense that they are manageable (i.e., not too extreme) and familiar (i.e., based on actual past events). However, although "realistic," the scenarios in such cases are also designed to constantly remind participants that they are not experiencing an actual reality. Thus, a clear line is drawn between actual reality and the simulated reality—the scenario—for example, by modeling a flu outbreak scenario as a "cat flu" event (which the participants clearly realize is not a real disease). Simulation exercise scenarios aim to help participants practice known solutions by simulating possible near-real present crises. They are not intended either to provoke thought about the future by creating something new or to make it possible to identify new problems by breaking with the known past.

In this regard, the role of the known past in simulation exercises is important and worth mentioning. Haunted by known past events such as SARS, pandemic flu, and Ebola, global public health actors have sought to develop and apply a range of different solutions. Of these, one of the most salient is that of the IHR. The IHR is thus a central document that participants in JADE and Crystal exercises are supposed to consult and become familiar with. During the exercise, participants are faced with an immediate crisis that can unfold in several possible ways. At each point in this unfolding crisis narrative, they are given a set of questions that are supposed to trigger a limited set of responses, often based on the IHR—for example, deciding whether to notify the WHO about an event in their country by using the decision instrument in Annex 2 of the IHR. In other words, these simulation exercise scenarios are designed to make participants practice known solutions to known past events.

Quite strikingly, in Exercise JADE, future uncertainty does not so much exist as a problem that is addressed with the simulation exercise as much as it is simply an intrinsic feature of the narrative. Indeed, participants have no influence over the development of the scenario and are not supposed to tackle or resolve any uncertainties they encounter in it. Recall that during the exercise, participants were not asked at any point to take actions that might constitute an actual investigation of the outbreak but simply had to respond to technical questions and to report on whether and how they had carried out the required procedures. Here, then, future uncertainty is merely structured in the narrative as an element that exists in actual events.

For example, at the beginning of the exercise, participants were told only that there had been reports of elderly people being hospitalized with a mysterious gastrointestinal infection whose cause was unknown (WHO EURO 2019b, Inject 1). For the participants, there could be multiple possible ways in which such a scenario might unfold. One newspaper article, for instance, suggested that the mysterious illness had spread via a mass-gathering event (WHO EURO 2019b, Inject 2)—a possibility that the participants, at that point, could neither completely rule out nor confirm. It was only on the seventh day in the scenario that participants received "official" confirmation that the mysterious illness had been caused by *Listeria monocytogenes*. As the story unfolded, then, initial uncertainties became certainties, and possibilities were narrowed down until a single, clear truth appeared. In this sense, while future uncertainty does exist in simulation exercises, it is structured by the technology and therefore predetermined and directed toward reassurance and certainty.

To summarize, in the field of global public health, the simulation exercise (as seen in the case of the WHO's Exercise JADE) is a central governmental technology that makes it possible to practice predetermined solutions, such as those set out in the IHR, that are largely driven by known past events. These solutions are practiced to enable improved performance in future events that are intrinsically unknown, unexpected, and uncertain. The simulation exercise is set to nearly mimic actual present reality and therefore includes an element of slowly diminishing future uncertainty (with developments becoming more certain as the scenario unfolds). Furthermore, and importantly, the simulation exercise generates a concrete experience of an immediate present crisis that provokes an urgent intervention, while it nevertheless creates a distinction between two spaces of intervention—between the actual present reality and the simulated present reality.

Security: Regoverning the Present through Plausible Futures

Turning Point exercises now form a key part of the national home-front preparedness agenda in Israel and have mainly taken the form of scenario-based exercises constructed around a war crisis event (and the associated crisis of civilian preparedness). Over the years, however, Turning Point exercises have also become rituals in their own right (see Samimian-Darash 2016; Samimian-Darash and Rotem 2019), as each exercise serves as a point of reference for the exercise the following year. That is, these exercises refer to the immediate occurrences of the past and the emergence of similar events that could develop into future crises.

Narratives for Turning Point scenarios include concrete and realistic events that draw both on information initially provided by Israeli intelligence agencies and on past and present events in Israel, either during real emergencies or from previous exercises. In addition, the narratives also include events that are yet to happen, as a way of breaking with the known past to create something new.

For example, in 2015, a war narrative was created, based on events and incidents that had taken place in the past as well as new imaginary threats and events that were devised for the sake of the exercise. The developed story was thus distinct from anything that had previously occurred but at the same time did not seek to predict what was yet to come. The recent past and the near future established the foundation for a narrative that was based on possible and plausible future events. That is to say, the time of the event was affected by known past events and the attributed future threat, thus developing events that were seen as potential but that had not actually taken place.

Since the goal of Turning Point exercises is neither to predict nor to realize the future before it happens, the exercise administration chooses a particular story, one that will form the basis of a plausible scenario, to practice that scenario in the present and see what might be learned about the future from the actual implementation of the exercise. Put differently, in these exercises, the time of the event, embedded in the complex scenario narrative, makes it possible to experience the uncertain future in the present.

This temporality is experienced during the actual implementation of the exercise, when real events from the past and fabricated future events are brought into a shared space of intervention. In this space, imagination is used to create a scenario in which the imagined narrative (events) and the "real" reality should not be viewed as mere fabulation or as truth, respectively. Rather, they are two coexisting realities: both are constructed and construct that which is real. As

Krasmann (2015, 188) argues, anticipatory knowledge practices always constitute a fictive reality that is distinct from the supposedly real reality but nonetheless real. This new reality, this imagined yet-to-be-realized narrative, is an important characteristic of the scenario, since it contributes to the real experience of the participants in the exercise as well as to keeping them involved and surprised every year and thus encourages them to act in the exercise as though it were a real present situation.

In these exercises, the broad virtual event presented in the scenario narrative is actualized and translated into practice through multiple incidents that have concrete repercussions. Many subevents emerge through this process, each with the potential to develop in unexpected ways depending on the reactions of participants and how those incidents interact with other incidents. Put differently, the prewritten narrative developed for the exercise shifts and changes as it adjusts to issues that emerge only during the actual practice of the exercise itself. Thus, while overall the exercise is a predesigned world, one of its most important aspects is that unpredictability arises at the level of the local actors and the interactions that take place between them. As a consequence of this temporality, participants constantly find themselves responding to unfolding incidents in ways that involve reactions to an emergent situation and subsequent actions upon it through organizational means and practices. However, since these actions themselves interact with the actions of other actors, the outcomes are also unpredictable and dynamic.

The Turning Point scenario addresses future uncertainty by creating plausible future stories and putting them into action in the present, not for the sake of prediction or making accurate prophecies but to create new knowledge about possible unknown futures. Keeping the scenario plausible depends on the provision of a story that is challenging yet not impossible to think of and prepare for. That is, these narratives are not worst-case scenarios; on the contrary, if they were too extreme, that would shift them from the real and reasonable to the unreal and impossible.

Practicing this format or these narratives (of plausible futures) enables the creation of a road map not only from the present to the future (from the past into prefuture predictions and assessments) but also from the future to the present, the result being that the imagined futures now govern the present (see, for example, the discussion of premediation in de Goede 2008a). Turning Point exercises enable thinking about the unknown future in the present to improve present actions toward it. In other words, instead of governing the future through other temporal orientations based on the past, the issue is one of governing the present by future models. That is, approaching the problem of emergent future uncertainty by creating imagined plausible future scenarios not only governs the future and reproblematizes the contemporary situation in the light of future orientation and concerns but also regoverns our present and makes the present an artifact or an outcome of future observations and actions.

Energy: Imagining Oneself in the Future

The World Energy Council's scenario planners create global energy scenarios that represent alternative plausible futures for the energy sector through a long process that involves various workshops with numerous representatives of central organizations in the energy field, including private companies, governmental ministries, and research institutions from around the world. These scenarios, it is claimed, provide "energy leaders with an open, transparent, and inclusive framework to think about the uncertain future, and thus assist in the shaping of the choices they make" (World Energy Council 2016, 2). The World Energy Council's scenarios seek to capture different futures resulting from a combination of predetermined elements and new critical uncertainties related to state policies and the market.

The representatives of the various energy organizations that take part in the World Energy Council's scenario-building events are first trained to accept uncertainty as a preliminary standpoint and then practice a process that involves narrative-building techniques, storytelling capabilities, systems thinking, and quantified illustrations, which together are used to create a framework for thinking about and addressing the future.

The narratives of the World Energy Council scenarios confront the problem of the uncertain future neither by making predictions nor by attempting to provide accurate assessments based on past information. Rather, they are plausible stories aimed at provoking a future-now thinking through which participants might think about possible drivers of the uncertain future and experience various future energy trends. Such a process does not set out to create ultimate knowledge about (possible) energy futures. Rather, the process represents the creation of a tool through which subjects can think about and act on the future as if they were in the future.

Throughout the process, participants are asked to think about how several different trends or themes might affect the three scenario narratives that were originally published in 2016 along with possible developments that might be expected in each scenario. Thus, workshops start with a focus on rethinking the previously created scenario narratives, while taking into account various possible developments in the future. The aim is to update the information and content contained in the 2016 narratives and to make changes that could point to more nuanced and up-to-date stories. Interestingly, the specificity, concreteness, and reasonableness of the scenario narratives are not about the taming of future uncertainty. These factors only assist in the creation of a good story that always, inherently, remains plausible (but not a possibility, in terms of past-based events) and subject to change. That is, the story reexpresses the idea that the potential future could always unfold in a different way.

The uncertainty rationality that is expressed in the narrative building of the scenarios is also linked to a particular temporality. Here, it is not simply that participants enact an imagined future to bring it to the present reality but rather that they (in the present) use imagination to bring themselves into a real future. In the World Energy Council workshops, participants are required to think about what comes from the future and to imagine that they are living in this "future present" and create "memories" of things that are yet to emerge. This is not to say that they are creating future presents—that is, that they realize the future yet to come. Rather, it is a way of making the scenario stories realistic. As facilitator and scenario expert Lorie Taylor explained to me during one of the workshops, "memories are pulled from the future; we are implementing [in the present story] what is imaginable," and this implementation provides the scenarios with their sense of realism. While the final versions of the scenarios that are produced during this process may resemble linear narratives that develop through consecutive chains of events, the process itself does not adhere to such a linear temporality.

The World Energy Council's scenario-planning process (chronological time) aims to tackle the problem of distant-future uncertainty, and the involvement of participants in that process is supposed to change how they think about the future and its relationship with the present. However, for such a change to occur, participants must recognize the importance of the distant future. Therefore, as they begin to use a "new language" and embrace a perception of the future as uncertain, heterogeneous, dynamic, and nonpredictable, they also come to accept the future and the present as equally important. In the workshops, present and near-past events are used together with stories about the future to build the scenarios (i.e., the time of event). Thereafter, participants are asked to think through these future scenarios to examine present events. Through this temporality, not only is the future realized in the present but the subjects imagine themselves in the future and create "new" futures.

Table 6.1 summarizes the dimensions of temporality in each of the three cases examined and indicates how a unique scenario-planning format is developed and expressed in each case, as well as the particular sense of what uncertainty is and how it is designed and expressed.

TABLE 6.1. Comparative temporal dimensions

FIELD	CHRONOLOGICAL TIME	TIME OF EVENT	TEMPORALITY OF INTERVENTION	UNCERTAINTY	SCENARIO-PLANNING FORMAT
	TEMPORAL CONTEXT OF THE FIELD	*TEMPORALITY OF THE NARRATIVE*	*TIMING AND AIM OF THE INTERVENTION*	*ORIENTATION TOWARD UNCERTAINTY*	*NARRATIVE AND PRACTICE*
WHO: Health	The known past and the currently occurring present	An immediate, urgent crisis in the present	Exercising/practicing specific, known solutions to past events toward the (possible) future	Structured uncertainty—an intrinsic feature of a crisis event	Storylines used to provoke a specific set of responses (simulation)
Turning Point: Security	The chronic emergency connecting the near past, present, and near future	A near future based on past information and new imagined events	Imagining and practicing the (possible and plausible) future to identify problems in the present	Uncertainty as emergence	An adaptable scenario narrative, which is both practiced and open for change and emergence during the exercise
World Energy Council: Energy	The distant future and potential future events	Potential futures related to the near past and the present	Thinking through and with multiple future events (as though one is living in the future) to think about unthinkable (plausible) futures	Uncertainty as in-becoming, potential and "future-present"	Participants are involved in the scenario design process (and imagining oneself in the future) and changed by it

While the WHO's simulation exercises emphasize the actual practice of particular responses and work toward automatically performing existing solutions in the immediate present (that is, an immediate present that will occur in the future), the World Energy Council's scenarios emphasize the narrative-building stage, acting toward improving future images and imagination as a form of experiencing and accepting uncertainty and future unexpected problems. In between, the Turning Point scenarios connect both imagination and practice, narrative building and exercising. The particular temporal orientation of these last exercises affects the format of the scenario-planning activities employed in this field, which are open for indetermination and directed toward an emerging uncertainty. Put differently, while scenario-planning activities at the WHO take the form of responses directed at exercising predetermined solutions, the Turning Point scenario-planning format focuses on the emergent and the detection of new problems that emerge in the exercise, and is thus used for further planning for the future. World Energy Council scenarios are not only about experiencing the future in the present but also about imagining oneself in the future.

While experts have looked for different possible solutions to the problem of future uncertainties either by translating uncertainty into risk conceptions and applying risk-based technologies toward it or by developing new uncertainty-based technologies to address it, in the anthropological analysis presented here I move from the issue of how to overcome the problem of the unknown future (a concern relevant to futurists) to a (social scientific) question regarding how future uncertainties themselves are a social product (i.e., culturally or socially created or selected). In addition, I inquire into how temporality, the different temporal models of each social context, affects future technologies and their particular orientation toward uncertainty. Hence, to understand the emergence of scenario-planning technologies and their increasing predominance, I suggest that it is necessary to consider how temporal perceptions affect this technology and its emergence and development in different sites.

That is, rather than analyzing uncertainty as merely an object of the future or a problem "in the world" that needs to be solved, or as a product of epistemology, a social perception or construct, in this chapter I have sought to emphasize how uncertainty is designed through the dynamic of each combination of scenario technique and temporality. What I have attempted to show is how the temporality of a particular approach to scenario planning affects the contours of uncertainty seen in each case. Furthermore, the different formats or configurations of scenario technologies re-create the ways in which we understand uncertainty, which now becomes not only an external future problem but also an internal product of the scenario technology. Uncertainty is thus designed, circulated, and created in different ways through the various formats of scenario technologies.

CONCLUSIONS AND CRITICAL LIMITATIONS

In a world of crises, pandemics, and unpredicted threats, use of the scenario technology has become widespread in efforts to prepare and plan for an uncertain future. This book provides a lens through which to view contemporary attempts to engage with future uncertainties via this technology. Taken together, the cases presented in the book enable us to understand in depth the methodology of future thinking and future making involved in the building of scenarios, how scenarios are put into action and facilitate preparedness exercises and practical responses, and how, in addition to expressing new modes of veridiction and jurisdiction, they also generate a particular mode of subjectivation. Overall, scenarios express a new modality of thinking, conceptualizing, imagining, exercising, and practicing future uncertainties in today's world—one that I term *uncertainty by design*.

I have suggested that scenarios emerged as a solution at a moment in time when a change was occurring in how the future was thought about and approached, and I have described how, under Herman Kahn's methodology of scenario thinking, concerns shifted from the problem of knowing the future to questions about the ways in which we think about it. Accepting that it was impossible to predict or accurately know uncertain futures, Kahn promoted the use of the imagination as a way of rendering unknown future events thinkable in the present. In Kahn's work, scenarios would thus be used as a speculative framework, a machine for generating what was hitherto unthinkable or unimaginable, to make it possible to prepare for such future possibilities in the present.

Moving on from Kahn's pioneering work, I have also shown how, through the subsequent work of scenario expert Pierre Wack, scenarios underwent a further shift in relation to uncertainty: from being a technology that accepts the uncertainty of the future as part of its reasoning to a technology that generates uncertainty in and through its practice. In Wack's work, scenarios went beyond accepting (an external future) uncertainty and working with it to become a way of producing a new perception—a means of encouraging managers to use uncertainty as an approach to decision making. Put differently, in Wack's approach, scenarios triggered a perception-changing process whose purpose was to make uncertainty a philosophical standpoint for seeing and acting in the world.

While scenarios emerged as a solution to the problem of providing knowledge about the unknown future—that is, facing an ontological problem concerning the nature of the future, which is unknown and uncertain—the scenario technology then shifted to address the epistemological question of how to think about the unknown future differently. Uncertainty has thus developed from being an ontological problem to being an epistemological premise. In this sense, the scenario technology has uncertainty as its mode of veridiction, jurisdiction, and subjectivation. That is, the scenario technology is based on uncertainty and expresses uncertainty in the knowledge it creates, the power of its exercises, and the subjective experience it generates.

I have also shown how scenario narrative-building activities work to create knowledge of the future (veridiction), and how the exercise of power (jurisdiction) in the actual implementation of the scenario-based exercise prompts more unknowns and uncertainties. Further, I have shown how the scenario-planning process changes subjective experience (subjectivation) to bring about a perception of the uncertain future in terms of potentiality, multiplicity, heterogeneity, and dynamism. Thus, I have reasserted that uncertainty is a rationality of governing that is distinct from risk and have argued that the scenario technology represents a new way of governing future uncertainty—uncertainty by design.

I have argued that governing future uncertainty (as expressed in the scenario technology) by embracing the emergent and the unexpected differs from governing future uncertainty through the rehearsal of predetermined solutions to a possible future event (as seen in use of the simulation technology). For this matter, scenarios and simulations differ in terms of both the construction of their narratives and the form of action and practice of the exercise itself—that is, the modes of veridiction and jurisdiction they incorporate.

I have also argued that the scenario technology simultaneously operates through two, nonlinear, temporal dimensions: the present future—narratives of an imagined future brought to and enacted in the present—and the future present—the "real" present that will take place in the future. Scenarios work on both the actual and the virtual while increasing and further producing uncertainty and potentiality. Further, I have discussed how particular temporalities in different fields influence the specific format of the scenario technology that is used.

Here it should be noted that while my project explores the scenario as a new modality of future governance, one that expresses a shift from probabilities and possibilities to imagination and potentiality, from risk-based technologies to uncertainty-based technologies, it is not my intention to idealize this shift. Nor is it my aim to advocate for the use of the scenario technology. While risk-based technologies convert reality into tamable possibilities and thus provide the appearance of control, the proliferation of the unexpected in the scenario also affects the reality of its participants and shapes broader societal perceptions of the future in today's world in particular ways.

Hence, while the rationality of uncertainty within the scenario technology promotes dynamism, potentiality, orientation toward the emergent, and problem extraction over solution-based mechanisms, it also expresses a constant tension between potentiality and the designing of futures as well as between accepting that the future is open-ended and impossible to control on the one hand and attempting to plan and prepare for it on the other. Put differently, like any other governmental technology, the scenario technology has its own externalities and critical limitations, to which I now turn.

In *Designing Human Practices*, Paul Rabinow and Gaymon Bennett (2012) distinguish between three modes of expert engagement in diverse problem spaces. Each of these three modes of engagement has its own externalities and critical limitations. If externalities are produced through a particular mode but are not taken explicitly into account, they can become "critical limitations—that is, they can introduce structural incapacities" (Rabinow and Bennett 2012, 52). Another way of putting this is to say that if externalities are what are excluded by a particular mode of thought, critical limitations are those things that cannot be thought through that mode.

I have already noted that the externalities of the scenario technology are connected to the prioritization of imagination over information and knowledge making. Scenarios are based on the acknowledgment of potential uncertainty that cannot be predicted, prevented, or calculated but that can be thought of and considered in efforts to narrate the future.

To illustrate the critical limitations of scenarios, I will discuss some aspects of the ongoing Coronavirus Disease 2019 (COVID-19) pandemic. I argue that the scenario technology's critical limitations are revealed when a shift takes place from future preparedness to responses to actual events. This leads to the idea that the critical limitations of scenarios are embedded in their temporality. Scenarios are created in the present toward real futures but act in a temporality of present future. When an event such as an unexpected pandemic or one based on an unknown cause occurs, we move from this present to the real future—that is, to the future present. In this reality, the problem of future uncertainty is replaced with a problem of planning for short-term future possibilities. Scenarios must then be updated or created anew (since they operate through the present future) or, alternatively, become descriptions of past possibilities (whether these had actualized or not) that cannot be used any longer as a guide to the new future present.

The COVID-19 Pandemic

In late December 2019, reports began to accumulate about a mysterious viral pneumonia outbreak in the Chinese city of Wuhan. On 31 December 2019, Wuhan health authorities reported that twenty-seven people (seven of them in serious condition), mostly people working at the Huanan Seafood Wholesale Market, had been admitted to the hospital with an unidentified respiratory illness. According to the reports, Chinese authorities began to conduct tests to identify the virus, commenced the implementation of public health measures (e.g., closure of the market), and notified the World Health Organization (WHO) to update it on the outbreak (Huang 2020; ProMED-mail 2019).

In the following days, the number of people diagnosed with the unknown pneumonia disease continued to rise, reaching forty-four by 3 January 2020, and Chinese authorities began to arrest people for spreading rumors about an outbreak of Severe Acute Respiratory Syndrome (SARS) on social media. Responding to this situation, countries in the local region that had imported cases of SARS from China in 2002–2003 began to take precautionary measures. Singapore began to apply temperature screening and isolation measures for all incoming travelers from Wuhan, and Hong Kong and Taiwan deployed thermal-imaging systems at boundary checkpoints (Gale 2020; ProMED-mail 2020a).

By 8 January, with fifty-nine people diagnosed in China, Chinese scientists managed to genetically sequence the virus and identified the cause of illness as a novel strain of coronavirus (Khan 2020).[1] As another week passed, it became clearer that the symptoms of the new virus included fever, malaise, dry cough, and shortness of breath. On 13 January, Thailand reported the first case outside China, a Chinese tourist who did not have any linkage to the Huanan Seafood Wholesale Market (Hui et al. 2020). By the end of January, the virus was continuing to spread across China and the region, and the number of diagnosed cases continued to rise. On 20 January, China reported a total of 217 cases (198 in Wuhan) and confirmed that the virus was transmissible between humans. Meanwhile, Japan and South Korea reported their first cases, and Thailand reported two cases. In response, the WHO called for an emergency meeting to assess whether the outbreak constituted a Public Health Emergency of International Concern (PHEIC) (Klar 2020; ProMED-mail 2020c). The WHO Emergency Committee convened on 22 January but did not recommend declaring a PHEIC, although its members agreed that the situation was urgent and that another meeting should be held in a matter of days. Meanwhile, multiple countries around the world began to test people who had recently returned from China for the coronavirus, setting up screening measures and imposing travel restrictions (ProMED-mail 2020d).

Just over a month later, by 3 March, the virus had managed to reach all six WHO regions (Western Pacific, South-East Asia, the Eastern Mediterranean, Africa, the Americas, and Europe), with significant local transmission occurring in South Korea (4,812 cases and 28 deaths), Italy (2,263 cases and 79 deaths), and Iran (2,336 cases and 77 deaths). By this stage, seventy-two countries (besides China) had been affected, with 90,869 confirmed cases globally (80,304 of them in China) and 166 deaths (in addition to 2,946 deaths in China) (ProMED-mail 2020e; WHO 2020a). On March 11, 2020, the WHO officially declared a pandemic (WHO 2020c).

By 18 April 2020, there were 2,160,207 confirmed cases and 146,088 deaths worldwide. Of these, 1,086,889 confirmed cases and 97,201 deaths were recorded in the WHO European Region alone (WHO 2020b). In these early months of the pandemic, as a growing number of governments around the world implemented strict public health measures and imposed nationwide lockdowns, a key problem that international and national authorities began to encounter was related to shortages in supplies of medical and protective equipment, especially ventilators and diagnostic tests (Rubin et al. 2020; Seipel 2020; Siddiqui 2020; WHO 2020s, 2).

In response to the international spread of the virus, the director-general of the WHO declared in late February that "every country needs to be ready to detect cases early, to isolate patients, trace contacts, provide quality clinical care, prevent hospital outbreaks, and prevent community transmission" (WHO 2020m). As ventilators and testing kits were crucial for countries' ability to perform these actions, they were in high demand and therefore in short supply. To help countries in this context, the WHO created the "COVID-19 Essential Supplies Forecasting Tool (ESFT)" (WHO 2020t). This tool, which has been occasionally updated in the light of new information, calculates input data to create forecasted estimates of needs in terms of essential supplies such as personal protective equipment, testing kits, oxygen therapy equipment, and essential drugs. With the move from the present to the real future, then, the problem of future uncertainty was replaced with the problem of how to plan for short-term future possibilities in accordance with the actualizing event.

In the following months, as COVID-19 resurged in many countries that had partially lifted public health restrictions, the capacity of health systems, hospitals, and intensive care units to withstand the pressure posed by surges of hospitalized cases had become a significant cause of worry (Gershengorn 2020; Stone and McMinn 2020; WHO AFRO 2020). Facing increasing hospitalizations and burnout among healthcare workers, healthcare systems became strained and overburdened, even in wealthy countries. Public health authorities thus warned that these systems might soon reach the brink of collapse. To support countries in their efforts to deal with such a situation, the WHO produced the "Adaptt Surge Planning Support Tool" (WHO EURO 2020a), which uses data input (e.g., from hospitals) to calculate and estimate possibilities of health-system incapacity—for instance, in relation to the required number of hospital beds. Another complementary tool is the "Health Workforce Estimator" (WHO EURO 2020b), which estimates the number of required health workers (of different kinds) and identifies future gaps in workforce capacity based on input and analysis of epidemiological data. These tools, then, address the specific problem of health systems' incapacity within the context of the broader problem of how to plan for short-term future possibilities.

By 3 March 2021, a bit over a year since its outbreak, 114.43 million confirmed cases of COVID-19 and 2.54 million deaths had been recorded worldwide. At this point, new concerns have emerged, particularly in the light of regulatory approvals of the first COVID-19 vaccines. These concerns are related to the ability of SARS-CoV-2 (the virus that causes COVID-19) to mutate, the increasing incidence of mutations, and, most importantly, the potential of the virus to mutate into new, more lethal or virulent variants. The emergence of these new concerns began with questions about the circulation of the virus between humans and animals when, in November 2020, Denmark reported on a cluster of human cases carrying a new variant of the virus that had circulated in mink populations (ProMED-mail 2020f). It then continued (from December 2020 onward) with even greater concerns due to reports on the emergence and rapid spread of new and highly transmissible variants in the United Kingdom and South Africa (ProMED-mail 2020g, 2020h). As increasing reports on the emergence of multiple new variants (e.g., in Bavaria and California) indicate, these developments have prompted substantial efforts to identify potentially dangerous mutations that could render existing vaccination efforts futile (Greaney et al. 2020; ProMED-mail 2020i, 2021a, 2021b, 2021c; WHO 2021; Zimmer 2021). Efforts to plan for future possibilities in the light of these concerns over the emergence of new and more lethal variants have prompted, for example, the creation of model-based projections for the spread of new variants (Galloway et al. 2021) and updates to forecasted estimates of morbidity and mortality in Europe (IHME 2021).

The COVID-19 pandemic presents a case of an actualization of potential uncertainty. Indeed, despite the accumulation of information and knowledge throughout the pandemic, every "wave" has brought with it more problems and new uncertainties. However, with the actualization of a potential unknown event, concerns shift to the immediate and urgent reality of the event—that is, to a temporality of a present reality that was previously set in the future. When this happens, the critical limitations of scenarios, which are most useful as imagination-based future narratives, are revealed.

From Potential Uncertainty to Pandemic

As the WHO and the experts advising it approached the emerging event, with all its uncertainties and unknowns, they did not imagine the potential ways in which the situation might unfold. Instead, they drew on past events to interpret the developing situation. This, however, proved challenging.

Following China's initial formal notification to the WHO China Country Office about the outbreak on 31 December 2019, from early January 2020 the WHO began to monitor the situation; developed, published, and continually updated technical guidance on the novel coronavirus; was regularly in contact with Chinese authorities; received updates from them about China's response efforts; and facilitated information sharing about the virus. An important part of the information sharing concerned genetic sequencing: "China shared the genetic sequence of the novel coronavirus on January 12, which will be of great importance for other countries to use in developing specific diagnostic kits" (WHO 2020d). In these efforts, the WHO played its role as an international body that coordinates between countries and facilitates information sharing. More so, however, the WHO sought to achieve greater clarity about the unknown virus—that is, an understanding of how the event would unfold. For this, the WHO needed information to ascertain what further information it would need to seek and how to do so on a continuous basis.

In early January, the WHO declared it could not yet assess the situation, since there was insufficient information to enable understanding of "the epidemiology, clinical picture, source, modes of transmission, and extent of infection; as well as the countermeasures implemented" (WHO 2020d). On the basis of the information available at that point, the organization concluded that it was not necessary to recommend "any specific health measures for travelers" and that "in case of symptoms suggestive of respiratory illness either during or after travel, travelers are encouraged to seek medical attention and share travel history with their healthcare provider" (WHO 2020d). In addition, the WHO advised "against the application of any travel or trade restrictions on China" (WHO 2020d).

Even as the virus continued to spread quickly across China, infected multiple healthcare workers, and reached other countries (WHO 2020g), these assessments and recommendations remained largely unchanged. This situation continued until late January. On 22 and 23 January, the WHO's director-general, Dr. Tedros Adhanom Ghebreyesus, convened an Emergency Committee (by teleconference) (WHO 2020h). Members of the Emergency Committee had different opinions on whether to declare a PHEIC. Of critical importance for the committee were evidence suggesting that human-to-human transmission had taken place, a preliminary R0 (a term used to signify how contagious an infectious disease is) estimate of 1.4–2.5, amplification of spread in healthcare settings (in at least one healthcare facility), and the fact that 25 percent of confirmed cases were severe. In addition, the source of the novel coronavirus remained unknown (though it was likely an animal reservoir), and there was no certainty regarding the extent of human-to-human transmission. Eventually, as some members argued it was too early to declare a PHEIC given the "restrictive and binary nature" of such a decision, the committee decided not to do so (though it was noted that this decision should be reexamined in a few days). Its conclusion was the following: "In the face of an evolving epidemiological situation and the restrictive binary nature of declaring a PHEIC or not, WHO should consider a more nuanced system, which would allow an intermediate level of alert. Such a system would better reflect the severity of an outbreak, its impact, and the required measures, and would facilitate improved international coordination, including research efforts for developing medical counter measures" (WHO 2020h).

Considering the information available at the time, the committee could not decide on whether the emergence of the novel coronavirus constituted a PHEIC. It was therefore felt that the existing system for declaring a PHEIC was insufficiently nuanced and had to change to enable a range of different, scaled decisions based on the continual appearance of new information.

The committee also recommended that the WHO assign an international multidisciplinary mission to support and review efforts to tackle the outbreak and offered recommendations for the "global community" based on previous experience with SARS (and the reforms of its operational procedures that had been implemented on the basis of the SARS outbreak): "As this is a new coronavirus, and it has been previously shown that similar coronaviruses required substantial efforts for regular information sharing and research, the global community should continue to demonstrate solidarity and cooperation, in compliance with Article 44 of the [International Health Regulations (2005)], in supporting each other on the identification of the source of this new virus, its full potential for human-to-human transmission, preparedness for potential importation of cases, and research for developing necessary treatment" (WHO 2020h).

Following the committee's advice, the WHO director-general did not declare a PHEIC. With regard to this decision, he remarked, "Make no mistake. This is an emergency in China, but it has not yet become a global health emergency. It may yet become one" (WHO 2020i). Elaborating on this, he reviewed what was known about the virus and then listed a number of aspects that remained unknown:

> We know that there is human-to-human transmission in China, but for now it appears limited to family groups and health workers caring for infected patients. At this time, there is no evidence of human-to-human transmission outside China, but that doesn't mean it won't happen. There is still a lot we don't know. We don't know the source of this virus, we don't understand how easily it spreads, and we don't fully understand its clinical features or severity. WHO is working with our partners . . . to fill the gaps in our knowledge as quickly as possible. It is likely that we will see more cases in other parts of China and other countries. (WHO 2020i)

Struggling with the uncertainty of a potential pandemic and the lack of knowledge about the novel coronavirus that might trigger it, the committee and the WHO drew on past events and existing solutions to comprehend and address the situation. Thus, the reality of the emerging event was constructed not only with new information but also with past-based knowledge.

The WHO then extrapolated this constructed reality into the future, converting potential uncertainty into possibilities by assessing the risk of the event as "very high" in China, "high" at the regional level, and "high" at the global level. In accordance with this assessment, the WHO undertook activities that included maintaining regular contact with affected countries; receiving information from these countries (in accordance with the International Health Regulations); informing other countries about the situation and providing support; activation of the incidents management system (at all WHO levels); development and updates of technical guidance, recommendations, and definitions; preparation of disease commodity packages containing supplies necessary for the identification and management of confirmed patients; utilization of global expert networks and partnerships for laboratory activities, infection prevention and control, clinical management, and mathematical modeling; activation of R&D

(research and development) blueprints to accelerate diagnostics, vaccines, and therapeutics; and joint coordination (with networks of researchers and experts) of global activities on surveillance, epidemiology, modeling, diagnostics, and clinical care and treatment (WHO 2020f).

On 30 January 2020, the Emergency Committee convened once again. This time, the committee advised the WHO to declare a PHEIC as local transmission outside China had been confirmed and the number of cases outside China was increasing (WHO 2020j). Explaining its decision, the committee "acknowledged that there are still many unknowns" and noted that "cases have now been reported in five WHO regions in one month, and human-to-human transmission has occurred outside Wuhan and outside China" (WHO 2020j). In regard to action, it stated that "the Committee believes that it is still possible to interrupt virus spread. . . . It is important to note that as the situation continues to evolve, so will the strategic goals and measures to prevent and reduce spread of the infection" (WHO 2020j).

Despite the decision to declare a PHEIC, then, the situation continued to evolve. This meant that determining what actions were and would be needed was highly dependent on more information becoming available from time to time. Thus, the committee recommended that actions be taken to enhance surveillance in China and to study the possible source of the virus, rule out hidden transmission, and inform risk management. Despite these recommendations, several weeks passed before the WHO changed its coronavirus threat assessment at the global level from "high" to "very high," its highest level of alert and risk assessment (WHO 2020k). What had driven this change was not a methodological change or new information but the way in which potential uncertainty is addressed.

At the end of February 2020, Mike Ryan, executive director of WHO's Health Emergencies Programme, responded to a question regarding whether the situation warranted the declaration of a pandemic: "If this was influenza we would probably have called this as a pandemic by now. But what we've seen with this virus is that with containment measures, with robust public health response the course of this epidemic or these multiple epidemics can be significantly altered" (WHO 2020n).

Observed in the light of past SARS outbreaks and possible influenza pandemics, the event was not declared a pandemic because it was seen as containable. There was still an opportunity to contain the virus and, Ryan continued, using the term *pandemic* would be "unhelpful" when efforts to contain the virus were still ongoing, because this would mean "giving up on the possibility of containing and slowing down the virus" and moving to "a phase of mitigation" (WHO 2020n). What was needed to avert a global pandemic, he claimed, was action: "If we don't take action, if we don't move, if we don't prepare, if we don't get ready that may be a future that we have to experience and we have to endure. So much of the future of this epidemic is not in the hands of the virus. A lot of the future of this epidemic is in the hands of ourselves, and those countries who've taken control, who've taken responsibility have clearly shown that a lot can be done to stop this virus" (WHO 2020n).

Whether the future would be one of containment or endurance of the virus was thus perceived as controllable—"in our hands, not in the hands of the virus." On the same day, the WHO director-general provided an update on the WHO's current understanding of the situation and the actions required:

> What we see at the moment are linked epidemics of COVID-19 in several countries, but most cases can still be traced to known contacts or clusters of cases. We do not see evidence as yet that the virus is spreading freely in communities. As long as that's the case, we still have a chance of containing this virus, if robust action is taken to detect cases early, isolate and care for patients and trace contacts. As I said yesterday, there are different scenarios in different countries, and different scenarios within the same country. The key to containing this virus is to break the chains of transmission. (WHO 2020l)

From the WHO's perspective, the best course of action was to contain the virus. Seeing that the situation could unfold into multiple different scenarios, at this "decisive point" in time, the director-general said on 27 February (WHO 2020m), countries that were reporting their first cases of the virus had a "window of opportunity" to contain it. Indeed, he continued, "the epidemics in the Islamic Republic of Iran, Italy and the Republic of Korea demonstrate what this virus is capable of. But this virus is not influenza. With the right measures, it can be contained" (WHO 2020m). In contrast to what was known about influenza, evidence of the lack of widespread community transmission in China suggested this virus could be contained. The WHO therefore called on countries to be ready for their first cases and their first clusters, and for community transmission, no matter their status or how good their health systems might be.

The WHO thus used new information and comparisons to past events to interpret the situation and create scenarios to plan for. In doing so, however, the WHO no longer just converted potential uncertainty into possibilities but also attempted to control and reduce potential future uncertainty. In this context and amid concerns for "the alarming levels of spread and severity, and . . . the alarming levels of inaction" (WHO 2020c), on 11 March 2020 the WHO officially declared a pandemic both because "the number of cases of COVID-19 outside China has increased thirteen-fold, and the number of affected countries has tripled" and because the WHO expected "to see the number of cases, the number of deaths, and the number of affected countries climb even higher" (WHO 2020c). In this regard, the WHO director-general stated,

> Pandemic is not a word to use lightly or carelessly. It is a word that, if misused, can cause unreasonable fear, or unjustified acceptance that the fight is over, leading to unnecessary suffering and death. Describing the situation as a pandemic does not change WHO's assessment of the threat posed by this virus. It doesn't change what WHO is doing, and it doesn't change what countries should do. We have never before seen a pandemic sparked by a coronavirus. . . . And we have never before seen a pandemic that can be controlled, at the same time. (WHO 2020c)

The WHO director-general decided to use the term *pandemic* to describe the ongoing event but clearly emphasized that the term now carried a different meaning and no longer indicated that "the fight is over." For the WHO, the assessment of the situation had not changed, since it was still grounded in the same constructed reality. The definition of the event as a pandemic, therefore, was not a moment when the WHO finally came to realize the actualization of a potential pandemic. Rather, it was a moment when the definition of the term itself had to be revised to incorporate the notion that potential uncertainty can be controlled by preset measures.

Simulations

The attempt to control the emerging uncertainty of the pandemic has been vividly present in the WHO's simulation exercises. The WHO encouraged countries to start preparing for the spread of the virus from the early phases of the outbreak. On 23 January 2020, the Emergency Committee convened by the WHO recommended that countries should start preparing: "It is expected that further international exportation of cases may appear in any country. Thus, all countries should be prepared for containment, including active surveillance, early detection, isolation and case management, contact tracing and prevention of onward spread of 2019-nCoV infection, and to share full data with WHO" (WHO 2020h).

This message was also conveyed on a regional level. On 24 January, Director Carissa Etienne of the WHO's Regional Office for the Americas, known as the Pan American Health Organization (PAHO), called on countries in the region "to be prepared to detect early, isolate and care for patients infected with the new coronavirus, in case of receiving travelers from countries where there is ongoing transmission of novel coronavirus cases" (PAHO 2020).

As the virus continued to spread, the WHO called for global efforts to help preparedness in regions that were expected to need support. On 30 January, the

Emergency Committee convened again and suggested that "in line with the need for global solidarity . . . global coordinated effort is needed to enhance preparedness in other regions of the world that may need additional support for that" (WHO 2020j). The rationale was that low-income countries would be most severely affected by the novel virus. Accordingly, several simulation exercises in preparation for COVID-19 were conducted by national health authorities with support from the WHO. Namibia, for example, held a tabletop simulation exercise for COVID-19 on 17 March (WHO 2020q). In addition, on 5–6 March, Ethiopia held a tabletop exercise to review the "operation management process for a suspected case of COVID-19" (WHO 2020r), to review plans—ranging from communications plans to risk and media communications plans—and to confirm arrangements and procedures prior to and following the confirmation of a COVID-19 case (WHO 2020r).

To provide preparedness support for countries, the WHO's Department of Health Security and Preparedness developed and published "a generic COVID-19 tabletop exercise package" (WHO 2020o). This package was developed for countries' national health authorities and includes a participants' guide and a facilitators' guide, reference documents and technical guidance on COVID-19, and a PowerPoint presentation to facilitate the exercise and subsequent debriefing. The exercise is divided into two parts (in the schedule provided in the package, each part stretches over half a day) and involves a group discussion to enable information sharing, identification of interdependencies between health actors and other sectors, conducting a gap analysis using the WHO's COVID-19 Strategic Preparedness and Response Plan, and the creation of a national action plan (WHO 2020p). The exercise package not only includes a scenario based on the ongoing pandemic but also has been updated occasionally to accommodate new information that emerged throughout it.

Much like the simulation exercises presented in chapter 5 of this book, here, too, the exercise was designed to simulate an event to test responses. In this context, it should be noted that the exercise mainly aims to "examine and strengthen existing plans, procedures and capabilities to manage an imported case of 2019-nCov" (WHO 2020o). The exercise is also built on scripted injects "to make participants consider the impact of a potential health emergency on existing plans, procedures and capacities" (WHO 2020p). Another similarity between this exercise and the simulation exercises presented in chapter 5 is the universal aspect that makes it adaptable to different localities: "The package highlights clearly where some minor adapt[at]ions are needed to make the simulation country-specific and more relevant to the participants" (WHO 2020o). Nevertheless, in the case of the COVID-19 exercise, the final update to the simulated narrative does not reduce uncertainty by informing participants of how the event comes to an end but instead leaves the participants with a quickly deteriorating situation that is becoming more, not less, uncertain. While this uncertainty is structured as part of the narrative—that is, inherent in the reality of the simulation—it is nevertheless practiced so it might be controlled. Thus, the simulation remains focused on practicing known solutions, as evident in the questions that follow narrative updates—for example, "What actions would be triggered by this new event information?" (WHO 2020p). Put differently, while the possible scenario that is simulated mimics an emerging event that has no definite closure and is therefore uncertain, the exercise is intended to control and reduce this uncertainty.

Scenarios

While some have responded to the potential uncertainty of the emerging pandemic and the related temporal shift toward the future present by drawing on past events, others have drawn on present-future scenarios, even though these scenarios were created in the past and were not updated or re-created.

On 18 October 2019, Johns Hopkins University's Center for Health Security, together with the World Economic Forum and the Bill & Melinda Gates Foundation, conducted a tabletop exercise known as "Event 201." This exercise lasted for 3.5 hours and involved "a series of dramatic, scenario-based facilitated discussions, confronting difficult, true-to-life dilemmas associated with response to a hypothetical, but scientifically plausible, pandemic" (Johns Hopkins Center for Health Security 2020a).

The exercise brought together fifteen actors from the public and private sectors (global businesses, government officials, and public health personnel, mainly from the United States) to practice a potential scenario where public-private partnerships would be required in the response to a pandemic with a significant socioeconomic impact (Johns Hopkins Center for Health Security 2020a, 2020b). In this, the participants had to resolve "real-world policy and economic issues" (Johns Hopkins Center for Health Security 2020a). The rationale for this exercise and the pandemic scenario in it were presented as follows:

> In recent years, the world has seen a growing number of epidemic events, amounting to approximately 200 events annually. These events are increasing, and they are disruptive to health, economies, and society. Managing these events already strains global capacity, even absent a pandemic threat. Experts agree that it is only a matter of time before one of these epidemics becomes global—a pandemic with potentially

catastrophic consequences. A severe pandemic, which becomes "Event 201," would require reliable cooperation among several industries, national governments, and key international institutions. (Johns Hopkins Center for Health Security 2020b)

Practicing the problem of a potential global pandemic with severe social and economic consequences was followed by a solution based on the need to practice cooperation between businesses, governments, and international institutions. As part of the exercise, different instruments were used to stimulate participation: prerecorded news broadcasts and moderated discussions on different topics that were "carefully designed in a compelling narrative that educated the participants and the audience" (Johns Hopkins Center for Health Security 2020a).

In building the scenario for the exercise, the organizers drew on an existing, known pathogen, SARS, yet modeled the simulated pathogen as more transmissible in the community, with mildly symptomatic cases able to transmit. The scenario, then, focused on an outbreak of a novel zoonotic coronavirus that jumped from bats to pigs to people and then became transmissible from person to person, which eventually led to a pandemic (Johns Hopkins Center for Health Security 2020c). The general narrative of the event is described as follows:

> The disease starts in pig farms in Brazil, quietly and slowly at first, but then it starts to spread more rapidly in healthcare settings. When it starts to spread efficiently from person to person in the low-income, densely packed neighborhoods of some of the megacities in South America, the epidemic explodes. It is first exported by air travel to Portugal, the United States, and China and then to many other countries. Although at first some countries are able to control it, it continues to spread and be reintroduced, and eventually no country can maintain control. There is no possibility of a vaccine being available in the first year. There is a fictional antiviral drug that can help the sick but not significantly limit spread of the disease. (Johns Hopkins Center for Health Security 2020c)

Finally, the scenario presented the eventual future development of the pandemic: "The scenario ends at the 18-month point, with 65 million deaths. The pandemic is beginning to slow due to the decreasing number of susceptible people. The pandemic will continue at some rate until there is an effective vaccine or until 80–90% of the global population has been exposed. From that point on, it is likely to be an endemic childhood disease" (Johns Hopkins Center for Health Security 2020c).

Excluding some of the specific details (e.g., initial outbreak location), this scenario is remarkably similar to the actual events that have unfolded in subsequent months. Even the socioeconomic problems described in the narrative are familiar: "Since the whole human population is susceptible, during the initial months of the pandemic, the cumulative number of cases increases exponentially, doubling every week. And as the cases and deaths accumulate, the economic and societal consequences become increasingly severe" (Johns Hopkins Center for Health Security 2020c).

The premise of the exercise was not to predict or assess the possibility that a particular future might happen but rather to prepare for the unpredictable future by inventing a realistic but nonetheless imaginary story and practicing it. Notwithstanding, the uncanny resemblance of this pandemic scenario to the COVID-19 pandemic has provided fertile ground for various interpretations, including conspiracy theories that have spread on the Internet, treating this (past-future) scenario as a prediction or, even worse, as a plan for the future. In response, the Johns Hopkins Center for Health Security released an announcement. In this, the center first declined that it had predicted the future: "To be clear, the Center for Health Security and partners did not make a prediction during our tabletop exercise. For the scenario, we modeled a fictional coronavirus pandemic, but we explicitly stated that it was not a prediction. Instead, the exercise served to highlight preparedness and response challenges that would likely arise in a very severe pandemic" (Johns Hopkins Center for Health Security 2020d).

Moreover, the announcement declared, the epidemiologic inputs at the basis of the model it had used to simulate the consequences of the virus in the scenario were different from those seen in the actual 2020 coronavirus pandemic: "We are not now predicting that the nCoV-2019 outbreak will kill 65 million people. Although our tabletop exercise included a mock novel coronavirus, the inputs we used for modeling the potential impact of that fictional virus are not similar to nCoV-2019" (Johns Hopkins Center for Health Security 2020d).

In other words, the center tried to correct people's wrong interpretation of the scenario as a past story that predicts the future. It insisted that it was not the scenario narrative itself that could be used as a representation for the presently occurring event but rather the new challenges that emerged from the practice of the scenario as a whole, in the context of the exercise. Thus, although scenarios appear to be relevant in preparedness for potential uncertain events, the temporal shift that accompanies an actualizing event renders scenarios less effective in responses that focus on an immediate and urgent future-present reality. Attempts to use scenarios in such situations highlight their critical limitations.

The critical limitations of the scenario technology emerge in the actualization of the event, not because scenarios are not relevant for the actual but because the actual is mistakenly treated as mere realization of a known possibility or misinterpreted with a past event rather than understood as an actualization of a potential uncertainty.

Reperceiving the Future

Scenarios emerge at the limits of the ability of present information to aid in efforts to prepare for the unknown future. Their future narratives are created neither as a prediction tool nor as a set of assessable possibilities (based on past information); instead, they shift knowledge making to imagination processes through which one can create plausible future stories, envisage new future complexities, and draw multiple pathways back to the present.

In analyzing how scenarios have been used in the case of the ongoing COVID-19 pandemic, one can see that scenario narratives are sometimes treated or perceived as an end point or as a way of creating one when the events they address materialize. That is, scenario narratives are treated as possibilities or predictions of the future and thus as diagrams of the present. The scenario as such is not a tool that can be used to prepare for potential uncertainty through processes of imagination (where the future is seen as something that is dynamic and inherently emerging and unfinished) but an image or a picture, a particular and possible representation of the future.

When crisis happens, when potential uncertainty is actualized, past events and plausible stories that were invented as a way of imagining the future are sometimes turned into possibilities or seen as a realization of the future. Instead of acknowledging the fact that these stories were not possibilities in the first place, people seek to choose the right scenario, the right story. The critical limitations of the scenario emerge at the occurring event not because scenarios are not relevant for the actual but because the actual is mistakenly treated as a realization of a present reality, a closed, known possibility. The possible takes over the plausible, and the real takes over the actual.

Furthermore, when the scenario is used merely as an image and not as a methodological tool for the imagination, this changes the nature of the scenario itself. The scenario is a *future* imagination (future-now thinking), not a map of the present written in the past about the past future. It is a future-imagining process rather than an image of a future used as a map of the present (which was created or drawn up in advance). Put differently, when one moves from preparing for future uncertainties to responding to an emerging crisis, the scenario must either change or become ineffective. The temporality of the scenario becomes its limitation.

If one wishes to use scenarios in the present, at a time of actualization, it is necessary to treat the present as a source of future potentiality and uncertainty rather than as the realization of a scenario that was written in the past. In other words, the temporality of the scenario cannot be shifted backward. The scenario is a tool created about the future to assist efforts to prepare for that future in the present, not a tool created in the past to enable action on the present. The shift of temporality involved here is not minor. Indeed, it goes to the very heart of the matter of the difference between risk-based thinking and uncertainty-based thinking.

"Scenarios give managers something very precious: the ability to re-perceive reality," argued Pierre Wack (1985b, 150). This is the ability to direct action and generate understanding in situations of uncertainty. The scenario approach aims to assist strategists and policy makers, who can use scenarios to "make sense" of their observations and experience, and thus have the "possibility to reframe their situation by considering alternative frames" and subsequently to "*reperceive* how their world works, the situation they are in, and the options that are available— and indeed, to generate new options—for action" (Ramírez and Wilkinson 2016, xii–xv, emphasis added). When used in this way, not only can scenarios enable better understanding through reperception, but they can also open up new possibilities and ways forward with action.

The scenario technology, through its dimensions of narration, exercising, and subjectivation as well as its unique temporality, thus provides a way to think of and act on future uncertainties by acknowledging their potentialities and promoting uncertainty as a rationality of practice.

EPILOGUE

Scenarios and the Dynamics between Science and Imagination

Imagination, wrote Vincent Crapanzano (2004, 19), "is much more than a faculty for evoking images which double the world of our direct perceptions: it is a distancing power thanks to which we represent to ourselves distant objects and we distance ourselves from present realities. . . . It permits fiction, the game, a dream, more or less voluntary error, pure fascination. It lightens our existence by transporting us into the region of the phantasm."

If indeed imagination reflects a horizon of openness and uncertainty, and practices of imagination increase possible futures and multiple subjective visions, how has this played out within modern scientific rationalities, perceptions, and methods?

The relationship between imagination and science is often framed by positioning imagination within art and as external to science. Historically speaking, however, such a framing is relatively recent. Indeed, imagination was a prominent concept in the science of the Romantic period of eighteenth-century Europe (Kearney 1988; Richardson 2013; Thomas 1999). At that time, the use of speculation, or imagination, in generating new theories constituted an integral part of scientific thinking (Sha 2013). Nevertheless, by the nineteenth century, imagination had come to be viewed as exterior to or even incompatible with science (Daston 1998; Huppauf and Wulf 2013, 3). That said, around the mid-twentieth century, in the wake of philosophical-historical discussions of science, the role of imagination in science was rediscovered, and scientific discourse has since witnessed a resurgence of interest in it (Salis and Frigg forthcoming; Stuart 2017). This resurgence of imagination within scientific discourse and practice, however, was still associated with art. It thus appears that imagination traveled back to science as an "art element" and that art and science were still assumed to be distinct forms with distinct practices.

In the context of scenario planning, imagination was a critical element in its emergence in the mid-twentieth century and was perceived as a solution to problems that could not be solved with the scientific tools available at the time. However, over the years, as various scenario-planning techniques and approaches were developed, attempts were made to increase scenario planning's scientific legitimacy. As a result, scenario methods have been reconstructed in terms of psychocognitive, behavioral, and quantitative scientific discourses and embedded in modern scientific rationale, which has led to imagination being concealed, removed, or made redundant.

However, an analysis of the work of contemporary scenario-planning experts could express an alternative dynamic between science and imagination. In this work, imagination is not restricted to art, nor is it limited to generating new theories. Rather, it appears as a central part or methodology in future planning, a practical scientific element of science. That is to say, I argue that imagination takes a new form and is used as a legitimate part of the "scientific research programme" (Lakatos 1980).

Science and Imagination between Romanticism and the Enlightenment

During the Romantic period of eighteenth-century Europe, scientists were less concerned with empiricism than with thought and ideas (Kearney 1988; Richardson 2013; Thomas 1999). At that time, "empiricism had the aura of mere manual and mechanic labor while theory had prestige. . . . That is to say, in order for scientists to have prestige, they had to distance themselves from mere mechanical labor" (Sha 2009, 663). Accordingly, the use of speculation, or imagination, in generating new theories was part of what being "scientific" meant (Sha 2013). Imagination, or speculation, was ontological, and "the real" perceived as resided in the realm of ideas rather than that of "things" (or empiricism) (Sha 2009, 663). Like "reality," "objectivity" had first been located in the "consciousness" or "the mind," but toward the end of the eighteenth century, as theory and speculation lost their prestige, these terms had come to be used to refer to objects external to consciousness, that "exist" in the world.

The emergence of science as "objective" (Daston and Galison 1992; Porter 1996) eventually led to the exclusion of imagination from its practice (Daston 1998). Scholars of the Enlightenment initially attempted to discipline imagination by applying the force of reason to it. Accordingly, both artists and scientists were seen as using a form of "restrained" imagination, unlike "the wild imagination that tyrannized pregnant women, religious fanatics, or mesmerized *convulsionnaires*" (Daston 1998, 88). Nevertheless, as science became more strictly factual and empirical, scientists grew more anxious and fearful about imagination's potential to threaten the boundaries between reality and fantasy (Daston 1998, 87–88; Sha 2009, 666).

Thus, the emergence of a particular type of mechanical objectivity in the midnineteenth century, a type that by the twentieth century had become a moral standard in scientific conduct (Daston and Galison 1992), enhanced the exclusion of imagination from science. Mechanical objectivity, Porter (1996, 4) asserts, "has been a favorite of positivist philosophers, and it has a powerful appeal to the wider public. It implies personal restraint. It means following the rules. Rules are a check on subjectivity: they should make it impossible for personal biases or preferences to affect the outcome of an investigation."

The Enlightenment fear of open-ended progress and the pursuit of mechanical objectivity that stemmed from this fear eventually led to scientists' growing emphasis on facts, and thus "pure facts, severed from theory and sheltered from the imagination, were the last, best hope for permanence in scientific achievement" (Daston 1998, 91). This development was consistent with the increasing "polarization of the personae of artist and scientist" and the relocation of imagination in the arts, outside science (Daston 1998, 74). The increasing association of science with empirical fact, the distancing of scientific from artistic work, and the increasing association of art with imagination helped solidify the notion of science as objective and exterior to consciousness, and of art as subjective and speculative: "Successful art could and did emulate scientific standards of truth to nature, and successful science could emulate artistic standards of imaginative beauty. But whereas in the eighteenth century both artists and scientists had seen no conflict in embracing both standards simultaneously, the chasm that had opened up between the categories of objectivity and subjectivity in the middle decades of the nineteenth century . . . forced an either/or choice" (Daston 1998, 86).

In short, the rise of mechanical objectivity in science led to imagination, seen in the Romantic period as scientific, or at least as a bridge between science and art (Sha 2009, 663), being viewed by the late nineteenth century as exterior to or even incompatible with science (Huppauf and Wulf 2013, 3).

Scenarios and Imagination

As discussed in chapter 1, scenario thinking first emerged in the early 1960s in response to the need to prepare for the possibility of nuclear war. Replacing the question of what the future will be with that of the form most suitable for thinking about the future, Herman Kahn problematized the way the future was approached and sought to address this problem through the use of scenarios. For Kahn, scenarios had to rely to some extent on reality and facts, but at the same time they needed to include the use of "vibrant imagination" in creating possible futures (Aligica 2004, 80). By imagining scenarios, Kahn maintained, it is possible to provoke thought about "unthinkable" futures. As he further asserted, imagination renders the unthinkable thinkable (Kahn 1984, 17). It enables one to consider possibilities that go beyond what we know historically or presently and is essential for thinking about the unknown future (Kahn 1960, 766–767). In Kahn's work, imagination was embraced. However, it was presented as distinct from science and existing scientific methods, as an external capacity that some individuals and scientists could express and use in their work.

As further discussed in chapter 1, scenario planning migrated into various think-tanks and research institutes, as well as to the business domain in the late 1960s and early 1970s. As a result, and especially following the success of Pierre Wack's scenario technique at Shell, scenarios soon gained prominence in the business world and received considerable attention from experts and scholars in fields such as strategy and planning (Chermack 2011). Nevertheless, this was also accompanied by a tendency to remove or underemphasize the scenario technique's imaginative element, often in an attempt to confer more scientific legitimacy on the practice.

In the business world, scenario experts had to justify the benefit and utility of the scenario technique to their managers while also defending the practice from criticism voiced by various experts in strategy and planning. In a much-cited article in the *Harvard Business Review*, Pierre Wack did precisely that in presenting the Shell scenario method.[1] He explained that this method could be used not only to analyze reality but also to change managers' perceptions of it, thus helping them to understand uncertainty (Wack 1985a, 76). In his publications, Wack discussed the development of the practice during his work for Shell, incorporating psychobiological justifications and anecdotes into his discussion to support his arguments (see, for example, Wack 1985a, 89). Wack framed scenarios as "scientific," using terms and knowledge drawn from the psychology of decision making. For example, his concepts of "macrocosm," "microcosm," "mental map," and "mental model" are interchangeable with psychological constructs such as "world view" and "mental image." Nevertheless, Wack still believed in the importance of imagination in scenario planning.[2]

Later practitioners followed Wack's justifications for the practice and tended to reduce its imaginative element. To make scenario planning more scientific, they refined the scenario technique and justified its value for decision making and psychology. These trends have been mostly apparent in the practitioner-based literature on scenario planning. One expression of this trend is a tendency to explain imagination in terms of psychological effect or cognitive structures—that is, to employ knowledge from psychology to explain functions involving the imaginative capacity.[3] This tendency can be seen, for instance, in Ronald Bradfield's (2008) study of cognitive learning barriers in the development of scenarios, in Fiona Graetz's (2002) study of left-brain and right-brain thinking styles in scenario planning, and in Alberto Franco et al.'s (2013) theoretical cognitive-style framework for studying individual differences in scenario planning.

Another expression of this trend (often overlapping with the first) has been the tendency to emphasize and develop scenarios for decision making and to overlook the role of imagination in them—that is, to overemphasize what scenario planning produces (without considering in full the role played by imagination) and to underemphasize how this is achieved. The hypothesis that scenarios have a psychological-behavioral impact on decision making has been elaborated by various experts since Wack first proposed this justification in his *Harvard Business Review* article. For example, Thomas J. Chermack and Kim Nimon (2008) conducted an experiment to test whether participation in scenario planning causes changes in decision-making styles, and George Wright and Paul Goodwin (2009) examined how scenario planning can mitigate the causes of low predictability in assessing low-probability events and how it might assist in resolving psychological issues such as inappropriate framing, cognitive and motivational bias, and inappropriate attributions of causality.[4] These attempts to improve, modify, or justify scenario planning to make it a more scientifically appreciated practice, then, appear to reduce its imaginative element to psychological-cognitive functions. Especially important here is contemporary scenario expert and academic Thomas Chermack, who has made major contributions to the effort to establish a scientific framework for the study of scenario planning, providing a comprehensive review, analysis, and synthesis of the literature on the subject (Chermack et al. 2001); examining the various definitions of scenario planning, analyzing dependent variables involved in the process, and suggesting an integrative definition for it (Chermack and Lynham 2002); proposing a theoretical model for scenario planning (Chermack 2004); and building a guide (which includes a project worksheet) on how to develop and use scenarios and how to assess the results of scenario projects (Chermack 2011).

Other experts have attempted to make the practice more "scientific" by actually attempting to fuse it with quantitative or calculative tools. The scenario approach associated with futurists—that is, those who view the future as technically predictable and use calculative models to know it (see Tolon 2011)—is a convergence between elements of scenario techniques and futurist forecasting, as developing scenarios and producing forecasts have become nearly indistinguishable (Bunn and Salo 1993, 292). In the application of futurist techniques such as trend-impact analysis, cross-impact analysis, and others, scenarios may be inserted at different points of the process. However, scenarios are only used within the boundaries set by the quantitative premise of futurist techniques—that is, they are aimed at producing knowledge about the future, whether this knowledge comes in accurate figures or probabilities (in determinist or indeterminist form, as Ian Hacking [1991] put it). In other words, in this approach, imagination was subordinated to scientific models (complex as they may be), and its role, if it had any at all, was restricted to the pregiven quantitative calculable framework, illustrated by or extended into agreed-on acts.

That said, the history of scenario planning suggests another option: the development of scenario planning as an imaginative-based scientific practice. A scenario expert who works with scenarios in various organizations once told me during an interview that "scenario planning is an art looking for science." By this he meant that there is a need to establish a scientific basis for the art of scenario planning—that is, a way for imagination itself (understood as an artistic capacity) to be perceived as scientific. In this sense, imagination as a way of reasoning—with scenario planning as the practice of this reasoning, so to speak—could be seen as a "science of imagination."

An Imaginative Scientific Practice

Scenario-planning practitioners have specifically emphasized the imaginative element of the scenario technique. Among those who have done so is Peter Schwartz, who succeeded Pierre Wack at Shell. Instead of subordinating scenario planning to the kind of science that aspires to accurately predicting the future or controlling biases in decision making, Schwartz has consistently endorsed imagination as a critical component of the practice. While he initially defined scenario design and implementation as an art (Schwartz 1996), he later suggested that it could be seen as both art and science: "While the approach is systematic, it does not pretend to the rigor of mathematical forecasting. Imagination and stretching conceptual boundaries are essential in developing good scenarios, hence the craft is as much an art as it is a science" (Schwartz 2002, 22).

This repositioning of scenario planning reemphasized the role of imagination in the process, and, more importantly, it emphasized imagination's value as an art and a science. In this regard, within this realm of science, the modern distinction between science and art, or science and imagination, falls apart.

Schwartz explained that the basic steps of scenario planning are as scientific as they are artful: "Typically, you will find yourself moving through the scenario process several times—refining a decision, performing more research, seeking out more key elements, trying on new plots, and rehearsing the implications yet again" (Schwartz 1996, 27–28). Further, when explaining the importance of producing different narratives as a way of communicating imagined futures in the scenario-building process, he implied that the art of scenario planning and science played equally important roles: "It is a common belief that serious information should appear in tables, graphs, numbers, or at least sober scholarly language. But important questions about the future are usually too complex or imprecise for the conventional languages of business and science. Instead, we use the language of stories and myths. Stories have a psychological impact that graphs and equations lack. . . . They help explain why things could happen in a certain way. They give order and meaning to events—a crucial aspect of understanding future possibilities" (Schwartz 1996, 37–38).

When one is dealing with the imprecise problem of the unknown future, that is, a subjectively imagined future, the scientific "language" of tables, graphs, and equations—external objects—is inadequate. Instead of justifying scenarios by appealing to modern scientific standards, that is, mechanical objectivity as a way of achieving intersubjective agreement, Schwartz justifies them in terms of their explanatory capacity in conditions where only subjective perceptions are available.

Another example demonstrates this point further. Former Shell scenario experts Rafael Ramírez and Angela Wilkinson (2016), together with another Shell veteran, Kees van der Heijden, developed the Oxford Scenario Planning Approach, which emphasizes the value of scenarios for strategic management and their use in a process of iteration, framing, and reframing. This, they claim, enables a mindset that is more open and flexible in conditions of TUNA (Turbulence, Uncertainty, Novelty, and Ambiguity). In other words, they justify scenarios as a means of achieving subjective understanding in situations where method and knowledge are ineffective or unavailable. They propose scenarios as a way of reperceiving and achieving understanding. A similar claim has been voiced by contemporary scenario expert Ged Davis, who has argued that "the use of images can help to make scenarios more comprehensible. Some aspects of scenarios may be described with numbers for use in the quantitative analysis of policies and strategy, but the richness of scenarios as a strategic tool stems partly from the fact that they can include more intangible aspects of the future" (Shell International 2003, 8).

For Davis, numbers and quantitative analysis can help communicate some aspects of scenarios, but an emphasis on stories and images can capture intangible aspects of the future. Similarly, as discussed in depth in chapter 4, the World Energy Council scenario team use quantitatively based diagrams and charts alongside stories to communicate the imagined forms of the future. Both cases suggest a new way of viewing imagination: not merely as an artistic capacity, the source of theories and hypotheses in science, or as a bridge between art and science. Imagination can be seen, instead, as a scientific practice in itself.

Scenarios and Imagination in the Twentieth Century and Beyond

Around the mid-twentieth century, in the wake of philosophical-historical discussions of science, the role of imagination in science was rediscovered, and scientific discourse has since witnessed a resurgence of interest in it (Salis and Frigg forthcoming; Stuart 2017). According to some accounts, science is "artful" and includes an imaginative element. For example, Warren Weaver (1929) argued that mathematics and physics (and hence modern science) are creative and imaginative arts. Similarly, Norman Campbell held that science and art (including the latter's imaginative element) are inseparable, and he pointed to nineteenth-century philosophy, with its anxiety over and concealment of the role of imagination in scientific discovery, as being responsible for the "ineptitude of modern scientific education" (Campbell 1957, 804). More particularly, Gerald Holton (1996) argued that for scientists to succeed, they must possess both the recognized tools of science (i.e., skills, logical reasoning, craftsmanship, and other disciplined expertise) and the tools of art (e.g., visual, analogical, and thematic imagination). However, when imagination—exiled from science throughout the nineteenth and early twentieth centuries—returned to scientific discourse and practice, it was still associated with art. It thus appears that imagination traveled back to science as an "art element" and that art and science were still assumed to be distinct forms or practices, as 1965 Nobel prizewinner François Jacob (2001, 118) maintained: "In the imaginative phase of scientific processes, in the formation of hypotheses, the scientist operates like an artist. It is only afterwards, when critical testing and experimentation are involved, that science draws away from art and goes down a different track."

Historical-philosophical studies of science have continued to reexamine the role of imagination in science. For instance, scholars have repositioned imagination as an elementary component of "thought experiments" carried out by scientists (Frappier et al. 2012; Gendler 2010; Mey 2006; Stuart 2017). Such experiments "bridge conceptual/theoretical knowledge to previous experience, existing knowledge and abilities" (Stuart 2017, 18). They produce understanding and assist in the development of new theories, models, and experiments through an imaginative exploration of the consequences of theoretical structures and interrelations between conceptualized phenomena (Stuart 2017, 12). Perhaps it is in the growing tendency within science to separate the theoretical and the empirical that imagination is returning to scientific discourse, as it is largely referenced in connection with the theoretical side.

However, imagination may interact with science on more than a theoretical basis: in the practice of scenario planning, there is an imaginative element that is both practical and (to an extent) empirical rather than merely theoretical. Here, imagination is not restricted to art, nor is it limited to generating new abstract theories. Accordingly, the historical analysis of scenario planning may indicate the emergence of an alternative conception of science, a novel "scientific research programme" (Lakatos 1980) in the pursuit of understanding at the limits of knowledge. Challenging views of science as disenchanting (Weber [1917] 2009), a socially trusted institutional source of counterfactually produced knowledge (Giddens 1990), a methodological search for truth through the rejection of theories and the accumulation of knowledge (Popper 1963), a type of modern ideology (Appadurai 1996), or a practice that is determined by dogmatic convictions held by communities or networks (Kuhn 1962; Margócsy 2017), the scientific research program apparent in contemporary scenario planning advances a different way of practicing science: One that does not seek to overcome subjective judgment by referring to external objects but builds on and evaluates quality with subjective judgment—with imagination. This entails focusing on dynamics, changes, and movements, thus offering new insights and options for action. This research program could then offer progress into new directions in scientific practice and its engagement with the world, as the practical use of scenario planning demonstrates: "In the end, it is decision and action that matter. But what does one do in the face of profound uncertainty? If one's belief is low, risks are low, and costs are high, then the primary initial action may simply be watching the situation. If you believe you can actually change the outcome in your favor and the risks are low, then affirmative action may be the right approach. If the belief in the scenario remains low but consequences high, then creating future options is the right track. Finally, if one believes in the discontinuity and there are large consequences then preemptive action may be warranted" (Schwartz 2002, 26–27).

As the previous quotation illustrates, scenario planning and the imaginative element in it offer a systematic and highly practical way of reasoning and deciding on action. The scientific research program it is a part of, then, advances a form of applied imagination-based scientific practice for societal issues. This example, however, represents just one interesting direction in which this scientific research program leads us: the ability to direct action and generate understanding in situations of uncertainty. Another quite similar direction is proposed by the Oxford Scenario Planning Approach. This approach aims to assist strategists and policy makers, who can use scenarios to "make sense" of their observations and experience and thus have the "possibility to reframe their situation by considering alternative frames" and subsequently to "*reperceive* how their world works, the situation they are in, and the options that are available—and indeed, to generate new options—for action" (Ramírez and Wilkinson 2016, xii–xv). In this direction of the scientific research program, as expressed by the Oxford Scenario Planning Approach, not only can imagination enable better understanding through reperception, but it can also open up possibilities and ways forward with action.

The work of scenario expert Adam Kahane on transformative scenario planning provides yet another instance of a direction in which this scientific research program can take us.[5] Kahane's experience with large-scale scenario projects, such as the "Mont Fleur" scenarios in South Africa and the "Destino Colombia" scenarios, taught him that by bringing together different people to imagine what *could* happen (rather than what they want to happen), then translating these imaginations into future scenarios, it is possible to transform participants' understandings, relationships, intentions, and actions and reshape transformative processes in the world (Kahane 2012, 17–18).[6]

For these experts, imagination is a necessary practice for increasing uncertainty, understanding change, practicing the future, and even promoting ways of shaping it (but not a means of knowing, controlling, or reducing future uncertainty). In this new embrace of imagination, then, imagination is seen as a legitimate part of a scientific practice. Such a view expresses the emergence of an alternative, and possibly new, scientific format that includes imagination at its core, as a rationale. Put differently, as practiced by some contemporary scenario experts, imagination implies a pursuit of a different kind of scientific research program, where progress is made by achieving understanding of the social world of which the researcher is a part. This, most importantly, marks an alternative view of science in which imagination is key to achieving understanding and making it possible to evaluate the quality of work.

To conclude, the history of scenario planning appears to express three axes in the problematization of the relationship between science and imagination: First, imagination as a nonscientific, exterior, artistic capability emerged as an alternative response for dealing with the problem of the unknown future, for which modern scientific thinking with its mechanical objectivity seemed inadequate. Second, imagination was perceived as pseudoscientific, and thus paradoxical attempts to make scenario planning more scientific led to the concealment or removal of imagination from it. Third, in a possibly larger shift in the dynamics of science and imagination, imagination has been once again embraced by scenario experts, this time as an independently justified scientific practice, as the "hard core" of a program that progressively develops and promotes this premise into various directions with different scenario-planning approaches.[7] As a result, what is defined or perceived as scientific is also changing. In this pursuit of novel understandings, the element of imagination does not make scenario planning pseudoscientific. In fact, it is the opposite that is the case. Imagination is accepted as a basis for scientific practices for researching the unknown future.

Notes

INTRODUCTION

1. I particularly looked into two cases: a smallpox vaccination project conducted in Israel in winter 2002–2003 and ongoing preparations for pandemic flu in 2005–2007 (see Samimian-Darash 2011a, 2011b). In preparing for the possibility of a future smallpox event, Israeli authorities were anticipating the recurrence of an event that had taken place in the past, and thus both the biological agent and the vaccine against it were known. Preparing for pandemic influenza invoked a different problematic, in which the threat was not possible but virtual. Because the virus that would be involved in a future outbreak was unknown, the disease's clinical case definition could not be ascertained in advance. Thus, not only did multiple types of influenza constitute a future virus's historical context, but they also provided the foundations for its clinical case definition. In other words, until an influenza pandemic event takes place, it is impossible to know what specific viral agent will be involved. In such a situation, an epidemic is not a fabrication or the semblance of an abstraction; rather, it is virtual, in that the potential for its appearance already exists, though this virtual occurrence can actualize as different events in the future—that is, as various pandemic strains that may require different types of treatment and response.

2. On the transformation of future unknowns into possibilities, that is, the "taming of chance," see Hacking (1990). In regard to the confined side of uncertainty-based technologies, Pat O'Malley (2015, 13) argues that according to the economic approach of Knight (1921) and Bernstein (1998), probabilistic prediction of the future is perceived as a "prison" that consigns us to an endless repetition of past statistical patterns over which we have no control. Uncertainty, on the contrary, is perceived as that which "makes us free." It is an expression of a set of cultural and philosophical beliefs that are integral to political liberalism.

3. See, for example, discussions of the "risk society" by Ulrich Beck (1992, 2009) and Anthony Giddens (1990).

4. For example, Mary Douglas and Aaron Wildavsky (1982) proposed that different dangers are selected for attention in different societies and thus come to be viewed as risks on the basis of particular cultural criteria .That is, there is "cultural bias" (Douglas 1994) of risk. Other scholars within science and technology studies have also discussed the link between risk and culture and in particular how scientific discourse "cleans" risk of its cultural and political origins. Sheila Jasanoff, for instance, argues that the scientific discourse on risk does this by "translating 'uncertainty' into formal quantitative language." She suggests that scholars should "restore the cultural dimension" by comparing "the discussion of uncertainty in different national settings" (Jasanoff 1999, 144). I similarly take as my premise that risks are socially constructed rather than given and hence depart from "rationalist" techno-scientific approaches to risk and uncertainty that aim to provide solutions, to mitigate or fix situations of risk and uncertainty, rather than to understand their social and cultural construction in the first place. However, as will be seen later, I also move from the cultural-epistemological argument to the governmental one.

5. See, for example, Anderson 2010a, 2010b; Aradau and van Munster 2007, 2011; Brassett and Vaughan-Williams 2015; Collier 2008; Lakoff 2008; Lentzos and Rose 2009; Schoch-Spana 2004.

6. Scenario experts usually use the term *scenarios* (plural) since, for them, the premise is that the future is open, and they therefore create multiple future scenarios. I use the singular term *the scenario*, however, since I discuss the approach as a governmental technology. In addition, a few anthropologies have offered to utilize scenarios as a tool for anthropological (Hannerz 2016; Heemskerk 2003).

7. In general, there is no rule regarding the "correct" number of scenarios, and it may change according to the purpose of a particular activity.

8. On extreme events, see, for example, Nassim Nicholas Taleb's discussion of "Black Swans": "Black Swans being unpredictable, we need to adjust to their existence (rather than naively try to predict them). There are so many things we can do if we focus on anti-knowledge, or what we do not know" (Taleb 2007, 3). On complex problems, see, for example, Brian Head's notion of "wicked problems," which are "complex, open-ended, and intractable." The term captures those situations in which there is no agreement over "the nature of the problem" or on its solution (Head 2008, 101), and such problems embed "clashes" between different values, structures, processes, and institutional arrangements (Head 2008, 104). Moreover, they have attributes that involve uniqueness and controversy over core issues such as problem definition and possible solutions (Rittel and Webber 1973). Such problems typically involve uncertainty stemming from lack of information, difficulty of articulation, and conflictual interactions (Koppenjan and Klijn 2004; Head 2008), as well as complexity and (value-based) divergence (Head 2008, 103). Moreover, they escape consensus with regard to what they are exactly and how they should be solved (Head 2008, 102).

9. In relation to psychology and decision making, see, for example, Kahneman and Tversky 1982; Runde 1998. In the context of management studies, see Gudykunst 1995. On uncertainty reduction theory, see Berger and Calabrese 1975.

10. Instead of asking whether bad or good uncertainty can better represent the future, however, I draw on Pat O'Malley's analysis to approach the risk-versus-uncertainty controversy from a different angle. Since risk and uncertainty offer different governmental forms, the question regarding the future is not what the future entails but, rather, how these technologies facilitate a dynamics either of imprisonment or of freedom (O'Malley 2004, 27) and how the future emerges as a problem of potential uncertainty to which these technologies are possible solutions.

11. See also Collier, Lakoff, and Rabinow 2004; Lakoff and Collier 2008; Rabinow 2003, 2007; Zinn 2009.

12. For Deleuze ([1968] 1994, 208–209), the virtual and the actual form two parts of the real object: "The virtual must be defined as strictly a part of the real object—as though the object had one part of itself in the virtual into which it plunged as though into an objective dimension. . . . There is another part of the object which is determined by actualization."

13. The story *Alice in Wonderland* provides a useful example to demonstrate the difference between the possible and the potential. As I explained elsewhere (Samimian-Darash 2013, 3n7), "when Alice stands in front of a road sign that points to two possible directions, the potentiality of her situation derives not only from her ability to choose one direction over the other (from 'freedom' of choice) but also from the form of the event itself, in which both paths exist simultaneously but neither actually exists until it is taken. As Deleuze ([1969] 1990, 3) puts it, there is no 'good sense' regarding the right direction (among multiple possibilities) and no 'commonsense as assignation of fixed identities'—the actualization of the virtual is not an action of representation of something that existed before. The path is actualized only when it is taken. There is no pre-existing 'event' or 'direction.'"

14. One solution suggested is "extended peer review" (Ravetz 2006), which involves broader scientific observations. One can see this solution as a call for second- and third-

order observations, pointing at each stage to blind spots that were not seen beforehand. In this regard, Luhmann (1998) argues that there are always blind spots, and thus no observation can "avoid" the risk inherent in a system's structure. Instead of "acknowledging" the problem of the immanent lack of knowledge, experts approach it by increasing communication in what Luhmann calls "the ecology of ignorance" (see also Rabinow 2007, 51–72).

15. Foucault ([2004] 2007) identifies three forms of governance: sovereignty, discipline, and biopolitical governmentality. Each emerged, he argues, in response to a specific governmental problem and was enacted to achieve a certain aim through determinate practices. Biopolitical security apparatuses emerged in response to the problem of circulation and freedom—that is, to address the need to regulate and secure the population. Security is thus seen as "a biopolitical problem of the protection and betterment of a population's essential life processes in an indeterminate world, rather than a geopolitical matter of prevention and exclusion" (Grove 2012, 140).

16. In this regard, Ian Hacking (1991) examines the production of statistical knowledge involved in creating a multiplicity of possibilities (i.e., the "emergence of probabilities"; see also Hacking 1975) and argues that these possibilities are then managed and controlled. As he puts it, "the erosion of determinism and the taming of chance by statistics does not introduce a new liberty.... The less determinism, the more the possibilities for constraint" (Hacking 1991, 194).

17. Michael Power (2004) describes the result of overusing risk management in modernity and particularly in organization management, which he terms "the risk management of everything." Risk is proliferated as related to organizational practices aimed at dealing with or controlling uncertainties. The problems rising from the growing expert accountability put on those performing risk analysis are a consequence of what he calls "the politics of uncertainty."

18. Another example of uncertainty-based technology can be seen in Adriana Petryna's work. According to Petryna (2015, 155), "horizoning work" is a new future technology that offers an alternative to risk-based technology: "horizons 'enfunction' uncertainty in a space of multiple parameters and provide an alternative heuristics to risk management, such as scenario planning, which have thus far failed to provide accurate depictions of risk or even clarity on decision making due to a variety of political and scientific constraints."

19. See also the *American Ethnologist*'s special collection "Orientations on the Future," available at https://americanethnologist.org/features/collections/orientations-to-the-future.

20. See also Samimian-Darash 2013.

21. Among these fields are those of policy (Volkery and Ribeiro 2009), energy (Benedict 2017), environmentalism (Wodak and Neale 2015), society (Ogilvy 2002), and crisis management (Moats et al. 2008).

22. For example, on precaution, preemption, and premediation, see Amoore 2013; de Goede 2008a, 2008b; de Goede and Randalls 2009; de Goede et al. 2014; Grusin 2004; Opitz 2017; Opitz and Tellmann 2015; Tellmann 2009; on resilience, see Brassett and Vaughan-Williams 2015; Lentzos and Rose 2009; on preparedness, see Aradau and van Munster 2012; Collier 2008; Lakoff 2007, 2008, 2017; Samimian-Darash 2016; and on anticipation, see Anderson 2010a, 2010b.

23. For literature in the context of security and emergency, see Adey and Anderson 2012; Aradau and van Munster 2011; Armstrong 2012; Brassett and Vaughan-Williams 2015; Collier 2008; Collier and Lakoff 2008; Cooper 2010; de Goede 2008a, 2008b; de Goede et al. 2014; Lakoff 2008; Masco 2008; Opitz and Tellmann 2015; Samimian-Darash 2016; Schoch-Spana 2004; Tellmann 2009. There are, however, some exceptions to this focus on security and emergency, with some studies directed more toward globalization (O'Brien 2016), law (Krasmann 2015), and environmentalism (Mathews and Barnes 2016). For literature on disastrous scenarios, see Aradau and van Munster 2011; Brassett and

Vaughan-Williams 2015; Collier 2008; Collier and Lakoff 2008; de Goede 2008a, 2008b; de Goede et al. 2014; Lakoff 2008; Opitz and Tellmann 2015; Schoch-Spana 2004; Tellmann 2009.

24. For literature that emphasizes the terms of unprecedented, unexpected, and uncertain, see Aradau and van Munster 2011; Cooper 2010; Krasmann 2015. For vulnerabilities, see Collier and Lakoff 2008; Lakoff 2007, 2008. About new knowledge and possible futures, see Aradau and van Munster 2011, 2012; Armstrong 2012; de Goede and Randalls 2009; Krasmann 2015; Tellmann 2009.

25. I focus here on the scenario approaches developed by Kahn and Wack, but it should be noted that a third origin or root that has been influential in the history of scenario planning is the so-called French School of scenario planning, also known as "La Prospective" (Godet 1986, 2000).

1. CHRONICITY

1. According to one of my informants, Pierre Wack and Ted Newland from Shell had engaged in meaningful and substantial discussion with Kahn and his colleagues from the Hudson Institute during the 1960s and 1970s.

2. Moltke, however, was notably more practice oriented than the more abstract and philosophical Clausewitz and, unlike the latter, never wrote a comprehensive piece articulating his ideas on warfare.

3. See also Amoore 2013.

4. Within the social sciences and humanities, scholars have studied the making, meaning, and effect of energy futures in relation to issues such as the global oil industry (Zalik 2010), geopolitics and energy security (Labban 2011), consulting firms' forecasts of the natural gas market (Mason 2007), visions of shale gas potential (Kuchler 2017), the making of unconventional fossil fuel resources (Kama 2020), sociotechnical imaginaries of energy systems in public policy (Jasanoff and Kim 2013), petroleum production and hydrocarbon prospects (Weszkalnys 2015), and the anticipation of economic disaster in the light of prospective oil exploration (Weszkalnys 2014).

2. NARRATIVE BUILDING

1. A rich body of work covers the topic of military and militarization in Israel (e.g., Ben-Eliezer 1998; Kimmerling 1993; Lomsky-Feder and Ben-Ari 2000; Sheffer and Barak 2010). Of interest and in reference to territorial borders, the sociologist Adriana Kemp (2000) argued that owing to historical political process, the very concept of borders grew obscured in the Israeli collective identity. The temporal border between war and peace is also fuzzy, as Shir-Vertesh and Markowitz (2015) noted. To them, the ongoingness of war in Israel and its entanglement with peace led to a temporal freezing point of "almost-peace/almost-war." Put differently, emergency and routine life melt into a state of "routinergency" (Shapiro and Bird-David 2017). Notably, Lebanese citizens residing on the other side of the border come to develop a similar attitude toward time and violence, a permanent mode of anticipation (Hermez 2017).

4. SUBJECTIVATION

1. I will not review all of the stages or the entire process of the workshops but only central parts that are relevant in terms of the analysis of scenario planning and subjectivity.

2. More recently Rabinow and Bennett (2012) refer to three "modes" (which they view as a more dynamic and contemporary term than *regimes*) of veridiction, jurisdiction, and subjectivation, which are mutually constituted and enable different forms of governance.

3. Foucault coins his own term for the constitution of subjectivity by using the term *subjectivation*, which he defined (in his last interview) as "the process by which one obtains the constitution of a subject, or more exactly, of a subjectivity, which is obviously only one of the given possibilities for organizing self-consciousness" (Foucault 2001, 1525, cited in Kelly 2010, 87; see also Foucault 1996, 472). He later even summarized his entire work as being related to the understanding of the relationship between the "subject" and the "truth."

5. SIMULATIONS

1. In 1995, with the rising threat of HIV/AIDS, spreading endemic diseases, and outbreaks of viral hemorrhagic fever, the World Health Assembly (WHA) decided that the WHO should revise its IHR. The revised 1995 version and the WHA resolutions adopted in 2001 and 2002 expanded the IHR to include activities related to early detection and rapid response to public health threats. Subsequently, growing international concerns over Severe Acute Respiratory Syndrome (SARS) and China's deliberate delayed notification to the WHO of the outbreak of that disease led to the WHA's adoption of a reformed version of the IHR (Gostin and Katz 2016, 266–267).

2. The Strategic Partnership for Health Security is a unit within the WHO's Country Health Emergency Preparedness & IHR (see https://extranet.who.int/sph/about-sph).

3. According to the IHR, each country shall establish a National IHR Focal Point, which is "the national centre, designated by each State Party, which shall be accessible at all times for communications with WHO IHR Contact Points under these Regulations" (WHO 2005, 8).

4. The EIS is an online secure portal accessible only to NFPs and selected international partners. The WHO uses it to disseminate information, risk assessments, and public health advice to NFPs.

5. But see also other circumstances, such as cases of smallpox or Severe Acute Respiratory Syndrome (SARS), that are unusual or unexpected and may have serious public health impacts.

7. CONCLUSIONS AND CRITICAL LIMITATIONS

1. Coronaviruses are enveloped, positive-sense single-stranded RNA genome viruses with helically symmetrical nucleocapsids. They have a spike (S) protein that enables their viral entry into target cells as the S-protein binds to a cellular receptor. Coronaviruses are part of the subfamily *Coronavirinae* in the family *Coronaviridae* (several members of this family constantly circulate in human populations, causing mild respiratory disease). In general, coronaviruses cause respiratory and gastrointestinal tract infections, and while various kinds of coronaviruses are known to exist, only some of them affect humans by causing illnesses, most importantly SARS and Middle East Respiratory Syndrome (MERS) (Hoffmann et al. 2020; ProMED-mail 2020b; Wu et al. 2020).

EPILOGUE

1. In the article, Wack (1985a, 74) asserted that managers recognized the value of scenarios—quoting, for instance, a former Shell director who had said that "experience has taught us that the scenario technique is much more conducive to forcing people to think about the future than the forecasting techniques we formerly used." Wack then went on to deal with skeptical experts in strategy and planning: "Many strategic planners claim they know all about scenarios: they have tried but do not like them." To these skeptics, he responded by arguing that the scenarios used at Shell were different from those that "merely quantify alternative outcomes of obvious uncertainties."

2. In a personal recorded conversation, Pierre Wack clearly stated that "creative imagination" was an important part of his work (with scenarios at Shell): "I know that in my life imagination is very important. On the good side and on the bad side. Sometimes I have ideas and, in my work, I use creative imagination without any doubt. And at other times, imagination is just dreaming and projecting things and not seeing things as they are" (cited in Chermack 2017, 19–20). It is clear from this conversation that Wack was aware of the importance of imagination in scenario planning, yet in his published work he barely mentioned imagination, let alone conceptually developed it.

3. This approach and its implications resonate with Alan Richardson's (2011) account of the current construction of imagination by cognitive scientists and neurologists.

4. These and other scenario experts constantly reflect on their practice and suggest new ways of using and improving scenario-planning practices—for instance, how to use scenarios to overcome inertia in strategic decision making (Wright et al. 2008) and how to improve scenario planning through the use of decision analysis to evaluate alternative strategies (Goodwin and Wright 2001).

5. Kahane, who joined Shell in 1988, learned the practice of scenario planning from scenario experts Ged Davis and Kees van der Heijden. Under their guidance, Kahane focused on understanding the world and changes in it as well as how to adapt to the latter. Nevertheless, influenced by the ideas of scenario expert Joseph Jaworski, he later began to focus on ways of "shaping" the future (Kahane 2012, 3).

6. The transformative scenario approach includes five basic steps. Generally, a team is convened from across the system (e.g., from political, social, and national spheres, depending on the context of the project) to observe a particular situation (including identification of its driving forces, certainties, and uncertainties). The team then comes up with stories that describe what could happen in that situation. As part of this exercise, they choose key certainties and uncertainties, construct scenarios, write logical narratives about hypothetical future events, use metaphors and pictures, compare and contrast the scenarios, and document the scenarios through various media. Next, the team discusses how to go forward in practical terms and, whether they choose an "adaptive" or a "transformative" approach, develop action plans based on possibilities and conclusions. Lastly, action is carried out to transform the system, maintaining the connections created during the project and promoting the realizations stemming from the scenarios (Kahane 2012, 97–98).

7. In other words, on the basis of Lakatos's view of what differentiates science from pseudoscience—whether a particular program's theories lead to the discovery of "hitherto unknown novel facts" (Lakatos 1980, 5)—we see how some contemporary scenario experts (e.g., Schwartz, Davis) are pursuing novel understandings, using various scenario approaches to progress toward this end.

References

Abram, Simone, and Gisa Weszkalnys. 2013. *Elusive Promises*. New York: Berghahn Books.

Adams, Vincanne, Michelle Murphy, and Adele E. Clarke. 2009. "Anticipation: Technoscience, Life, Affect, Temporality." *Subjectivity* 28(1): 246–265.

Adey, Peter, and Ben Anderson. 2012. "Anticipating Emergencies: Technologies of Preparedness and the Matter of Security." *Security Dialogue* 43(2): 99–117.

Ahmann, Chloe. 2018. "It's Exhausting to Create an Event out of Nothing": Slow Violence and the Manipulation of Time." *Cultural Anthropology* 33(1): 142–171.

Aligica, Paul Dragos. 2004. "The Challenge of the Future and the Institutionalization of Interdisciplinarity: Notes on Herman Kahn's Legacy." *Futures* 36(1): 67–83.

Amoore, Louise. 2013. *The Politics of Possibility: Risk and Security beyond Probability*. Durham, NC: Duke University Press.

Anderson, Ben. 2007. "Hope for Nanotechnology: Anticipatory Knowledge and the Governance of Affect." *Area* 39(2): 156–165.

Anderson, Ben. 2010a. "Preemption, Precaution, Preparedness: Anticipatory Action and Future Geographies." *Progress in Human Geography* 34(6): 777–798.

Anderson, Ben. 2010b. "Security and the Future: Anticipating the Event of Terror." *Geoforum* 41(2): 227–235.

Anderson, Ben, and Peter Adey. 2011. "Affect and Security: Exercising Emergency in 'UK Civil Contingencies.'" *Environment and Planning D: Society and Space* 29(6): 1092–1109.

Andersson, Jenny. 2018. *The Future of the World: Futurology, Futurists, and the Struggle for the Post–Cold War Imagination*. Oxford: Oxford University Press.

Antonello, Alessandro, and Mark Carey. 2017. "Ice Cores and the Temporalities of the Global Environment." *Environmental Humanities* 9(2): 181–203.

Appadurai, Arjun. 1996. *Modernity at Large: Cultural Dimensions of Globalization*. Minneapolis, MN: University of Minnesota Press.

Appadurai, Arjun. 2013. *The Future as Cultural Fact: Essays on the Global Condition*. London: Verso.

Aradau, Claudia, Luis Lobo-Guerrero, and Rens van Munster. 2008. "Security, Technologies of Risk, and the Political: Guest Editors' Introduction." *Security Dialogue* 39(2–3): 147–154.

Aradau, Claudia, and Rens van Munster. 2007. "Governing Terrorism through Risk: Taking Precautions, (Un)Knowing the Future." *European Journal of International Relations* 13(1): 89–115.

Aradau, Claudia, and Rens van Munster. 2011. *Politics of Catastrophe: Genealogies of the Unknown*. Abingdon, UK: Routledge.

Aradau, Claudia, and Rens van Munster. 2012. "The Securitization of Catastrophic Events: Trauma, Enactment, and Preparedness Exercises." *Alternatives: Global, Local, Political* 37(3): 227–239.

Armstrong, Melanie. 2012. "Rehearsing for the Plague: Citizens, Security, and Simulation." *Canadian Review of American Studies* 42(1): 105–120.

Bear, Laura. 2014. "Doubt, Conflict, Mediation: The Anthropology of Modern Time." *Journal of the Royal Anthropological Institute* 20: 3–30.

Bear, Laura. 2016. "Time as Technique." *Annual Review of Anthropology* 45: 487–502.

Beck, Ulrich. 1992. *Risk Society: Toward a New Modernity.* London: Sage.

Beck, Ulrich. 2009. *World at Risk.* Cambridge: Polity Press.

Ben-Eliezer, Uri. 1998. *The Making of Israeli Militarism.* Bloomington, IN: Indiana University Press.

Benedict, Barry A. 2017. "Benefits of Scenario Planning Applied to Energy Development." *Energy Procedia* 107: 304–308.

Berger, Charles R., and Richard J. Calabrese. 1975. "Some Explorations in Initial Interaction and Beyond: Toward a Developmental Theory of Interpersonal Communication." *Human Communication Research* 1(2): 99–112.

Bernstein, Peter L. 1998. *Against the Gods: The Remarkable Story of Risk.* New York: Wiley.

Beyerchen, Alan. 1992. "Clausewitz, Nonlinearity, and the Unpredictability of War." *International Security* 17(3): 59–90.

Bradfield, Ron M. 2008. "Cognitive Barriers in the Scenario Development Process." *Advances in Developing Human Resources* 10(2): 198–215.

Bradfield, Ron, George Wright, George Burt, George Cairns, and Kees van der Heijden. 2005. "The Origins and Evolution of Scenario Techniques in Long Range Business Planning." *Futures* 37(8): 795–812.

Brassett, James, and Nick Vaughan-Williams. 2015. "Security and the Performative Politics of Resilience: Critical Infrastructure Protection and Humanitarian Emergency Preparedness." *Security Dialogue* 46(1): 32–50.

Brigstocke, Julian. 2016. "Exhausted Futures." *GeoHumanities* 2(1): 92–101.

Bryant, Rebecca, and Daniel M. Knight. 2019. *The Anthropology of the Future.* Cambridge: Cambridge University Press.

Bunn, Derek W., and Ahti A. Salo. 1993. "Forecasting with Scenarios." *European Journal of Operational Research* 68(3): 291–303.

Campbell, Norman Robert. 1957. "Science, Imagination, and Art." *Science* 125(3252): 803–806.

Chandler, David, and Julian Reid. 2016. *The Neoliberal Subject: Resilience, Adaptation and Vulnerability.* London: Rowman & Littlefield.

Chermack, Thomas J. 2004. "A Theoretical Model of Scenario Planning." *Human Resource Development Review* 3(4): 301–325.

Chermack, Thomas J. 2011. *Scenario Planning in Organizations: How to Create, Use, and Assess Scenarios.* Oakland, CA: Berrett-Koehler.

Chermack, Thomas J. 2017. *Foundations of Scenario Planning: The Story of Pierre Wack.* New York: Routledge.

Chermack, Thomas J., and Laura M. Coons. 2015. "Scenario Planning: Pierre Wack's Hidden Messages." *Futures* 73: 187–193.

Chermack, Thomas J., and Susan A. Lynham. 2002. "Definitions and Outcome Variables of Scenario Planning." *Human Resource Development Review* 1(3): 366–383.

Chermack, Thomas J., Susan A. Lynham, and Wendy E. A. Ruona. 2001. "A Review of Scenario Planning Literature." *Futures Research Quarterly* 17(2): 7–32.

Chermack, Thomas J., and Kim Nimon. 2008. "The Effects of Scenario Planning on Participant Decision-Making Style." *Human Resource Development Quarterly* 19(4): 351–372.

Claeys, Gregory. 2010. "The Origins of Dystopia: Wells, Huxley and Orwell." In *The Cambridge Companion to Utopian Literature,* edited by Gregory Claeys, 107–134. New York: Cambridge University Press.

Clarke, Lee. 1999. *Mission Improbable: Using Fantasy Documents to Tame Disaster*. Chicago, IL: University of Chicago Press.

Clausewitz, Carl von. 1982. *On War*. London: Penguin.

Collier, Stephen J. 2008. "Enacting Catastrophe: Preparedness, Insurance, Budgetary Rationalization." *Economy and Society* 37(2): 224–250.

Collier, Stephen J., and Andrew Lakoff. 2008. "Distributed Preparedness: The Spatial Logic of Domestic Security in the United States." *Environment and Planning D: Society and Space* 26(1): 7–28.

Collier, Stephen J., Andrew Lakoff, and Paul Rabinow. 2004. "Biosecurity: Towards an Anthropology of the Contemporary." *Anthropology Today* 20(5): 3–7.

Cooper, Melinda. 2006. "Pre-empting Emergence: The Biological Turn in the War on Terror." *Theory, Culture and Society* 23(4): 113–135.

Cooper, Melinda. 2010. "Turbulent Worlds." *Theory, Culture and Society* 27(2–3): 167–190.

Crapanzano, Vincent. 2004. *Imaginative Horizons: An Essay in Literary-Philosophical Anthropology*. Chicago, IL: University of Chicago Press.

Cuzzocrea, Valentina, and Giuliana Mandich. 2016. "Students' Narratives of the Future: Imagined Mobilities as Forms of Youth Agency?" *Journal of Youth Studies* 19(4): 552–567.

D'Angelo, Lorenzo, and Robert J. Pijpers. 2018. "Mining Temporalities: An Overview." *The Extractive Industries and Society* 5(2): 215–222.

Daston, Lorraine. 1998. "Fear and Loathing of the Imagination in Science." *Daedalus* 127(1): 73–95.

Daston, Lorraine, and Peter Galison. 1992. "The Image of Objectivity." *Representations* 40: 81–128.

Davis, Ged. 2002. "Scenarios as a Tool for the 21st Century." Presentation at the Probing the Future Conference, Strathclyde University, Glasgow, 12 July.

Davis, Tracy C. 2007. *Stages of Emergency: Cold War Nuclear Civil Defense*. Durham, NC: Duke University Press.

de Goede, Marieke. 2008a. "Beyond Risk: Premediation and the Post-9/11 Security Imagination." *Security Dialogue* 39(2–3): 155–176.

de Goede, Marieke. 2008b. "The Politics of Preemption and the War on Terror in Europe." *European Journal of International Relations* 14(1): 161–185.

de Goede, Marieke, and Samuel Randalls. 2009. "Precaution, Preemption: Arts and Technologies of the Actionable Future." *Environment and Planning D: Society and Space* 27(5): 859–878.

de Goede, Marieke, Stephanie Simon, and Marijn Hoijtink. 2014. "Performing Preemption." *Security Dialogue* 45(5): 411–422.

Dean, Mitchell. 1999. *Governmentality: Power and Rule in Modern Society*. Cambridge: Cambridge University Press.

Deleuze, Gilles. [1969] 1990. *The Logic of Sense*. New York: Columbia University Press.

Deleuze, Gilles. [1968] 1994. *Difference and Repetition*. New York: Columbia University Press.

Deleuze, Gilles, and Félix Guattari. 1987. *A Thousand Plateaus: Capitalism and Schizophrenia*. Minneapolis, MN: University of Minnesota Press.

Deleuze, Gilles, and Félix Guattari. [1991] 1994. *What Is Philosophy?* New York: Columbia University Press.

Derrida, Jacques. 1988. "The Politics of Friendship." *Journal of Philosophy* 85(11): 632–644.

Diprose, Rosalyn, Niamh Stephenson, Catherine Mills, Kane Race, and Gay Hawkins. 2008. "Governing the Future: The Paradigm of Prudence in Political Technologies of Risk Management." *Security Dialogue* 39(2–3): 267–288.

Douglas, Mary. 1994. *Risk and Blame: Essays in Cultural Theory*. London: Routledge.

Douglas, Mary, and Aaron Wildavsky. 1982. *Risk and Culture: An Essay on the Selection of Technical and Environmental Dangers*. Berkeley, CA: University of California Press.

Ewald, François. 1991. "Insurance and Risks." In *The Foucault Effect: Studies in Governmentality*, edited by Graham Burchell, Colin Gordon, and Peter Miller, 197–210. Chicago, IL: University of Chicago Press.

Faubion, James D. 2019. "On Parabiopolitical Reason." *Anthropological Theory* 19(2): 219–237.

Fischer, Edward. 2014. *The Good Life: Aspiration, Dignity, and the Anthropology of Wellbeing*. Redwood, CA: Stanford University Press.

Foucault, Michel. 1990a. *The History of Sexuality: An Introduction*. Vol. 1. New York: Vintage.

Foucault, Michel. 1990b. *The History of Sexuality: The Use of Pleasure*. Vol. 2. New York: Vintage.

Foucault, Michel. 1991. "Questions of Method." In *The Foucault Effect: Studies in Governmentality*, edited by Graham Burchell, Colin Gordon, and Peter Miller, 73–86. Chicago, IL: University of Chicago Press.

Foucault, Michel. 1996. *Foucault Live*. Edited by Sylvère Lotringer. New York: Semiotext(e).

Foucault, Michel. 2001. *Dits et écrits*. Vol. 2. Edited by Daniel Defert and François Ewald. Paris: Gallimard.

Foucault, Michel. [2004] 2007. *Security, Territory, Population: Lectures at the Collège de France, 1977–78*. Translated by Graham Burchell. London: Palgrave.

Franco, L. Alberto, Maureen Meadows, and Steven J. Armstrong. 2013. "Exploring Individual Differences in Scenario Planning Workshops: A Cognitive Style Framework." *Technological Forecasting and Social Change* 80(4): 723–734.

Frappier, Melanie, Letitia Meynell, and James Robert Brown. 2012. *Thought Experiments in Science, Philosophy, and the Arts*. New York: Routledge.

Funtowicz, Silvio O., and Jerome R. Ravetz. 1991. "A New Scientific Methodology for Global Environmental Issues." In *Ecological Economics: The Science and Management of Sustainability*, edited by Robert Costanza, 137–152. New York: Columbia University Press.

Funtowicz, Silvio O., and Jerome R. Ravetz. 1993. "Science for the Post-Normal Age." *Futures* 25(7): 739–755.

Gale, Jason. 2020. "Pneumonia Outbreak in China Spurs Fever Checks from Singapore to Taiwan." *Bloomberg*, 3 January. Available at: https://www.bloomberg.com/news/articles/2020-01-03/pneumonia-outbreak-spurs-fever-checks-from-singapore-to-taiwan (accessed 22 March 2020).

Galison, Peter. 1996. "Computer Simulations and the Trading Zone." In *The Disunity of Science: Boundaries, Contexts, and Power*, edited by Peter Galison and David J. Stump, 118–157. Stanford, CA: Stanford University Press.

Galloway, Summer E., Paul Prabasaj, Duncan R. MacCannell, Michael A. Johansson, John T. Brooks, Adam MacNeil, Rachel B. Slayton, Suxiang Tong, Benjamin J. Silk, Gregory L. Armstrong, Matthew Biggerstaff, and Vivien G. Dugan. 2021. "Emergence of SARS-CoV-2 B.1.1.7 Lineage—United States, December 29, 2020–January 12, 2021." *Morbidity and Mortality Weekly Report (MMWR)*. Early Release, 15 January 2021. http://dx.doi.org/10.15585/mmwr.mm7003e2.

Gell, Alfred. 1992. *The Anthropology of Time: Cultural Constructions of Temporal Maps and Images*. Oxford: Berg.

Gendler, Tamar. 2010. *Intuition, Imagination, and Philosophical Methodology*. Oxford: Oxford University Press.

Gershengorn, Hayley. 2020. "ICU Capacity Is More about the Clinicians than the Number of Beds." *STAT*, 18 August. Available at: https://www.statnews.com/2020/08/18/icu-capacity-is-more-about-the-clinicians-than-the-number-of-beds/ (accessed 20 January 2021).

Ghamari-Tabrizi, Sharon. 2005. *The Worlds of Herman Kahn: The Intuitive Science of Thermonuclear War.* Cambridge, MA: Harvard University Press.

Giddens, Anthony. 1990. *The Consequences of Modernity.* Cambridge: Polity Press.

Godet, Michel. 1986. "Introduction to La Prospective: Seven Key Ideas and One Scenario Method." *Futures* 18(2): 134–157.

Godet, Michel. 2000. "The Art of Scenarios and Strategic Planning: Tools and Pitfalls." *Technological Forecasting and Social Change* 65(1): 3–22.

Goodwin, Paul, and George Wright. 2001. "Enhancing Strategy Evaluation in Scenario Planning: A Role for Decision Analysis." *Journal of Management Studies* 38(1): 1–16.

Gordon, Theodore Jay. 1994. "The Delphi Method." *Futures Research Methodology* 2(3): 1–30.

Gostin, Lawrence O., and Rebecca Katz. 2016. "The International Health Regulations: The Governing Framework for Global Health Security." *The Milbank Quarterly* 94(2): 264–313.

Graetz, Fiona. 2002. "Strategic Thinking versus Strategic Planning: Towards Understanding the Complementarities." *Management Decision* 40(5): 456–462.

Greaney, Allison J., Andrea N. Loes, Katharine H. D. Crawford, Tyler N. Starr, Keara D. Malone, Helen Y. Chu, and Jesse D. Bloom. 2020. "Comprehensive Mapping of Mutations to the SARS-CoV-2 Receptor-Binding Domain that Affect Recognition by Polyclonal Human Serum Antibodies." *bioRxiv* preprint, 4 January 2021. https://doi.org/10.1101/2020.12.31.425021.

Grove, Kevin. 2012. "Preempting the Next Disaster: Catastrophe Insurance and the Financialization of Disaster Management." *Security Dialogue* 43(2): 139–155.

Grusin, Richard. 2004. "Premediation." *Criticism* 46(1): 17–39.

Gudykunst, William D. 1995. "Anxiety/Uncertainty Management (AUM) Theory: Current Status." In *International and Intercultural Communication Annual*, Vol. 19, *Intercultural Communication Theory*, edited by Richard L. Wiseman, 8–58. Thousand Oaks, CA: Sage.

Hacking, Ian. 1975. *The Emergence of Probability.* Cambridge: Cambridge University Press.

Hacking, Ian. 1990. *The Taming of Chance.* Cambridge: Cambridge University Press.

Hacking, Ian. 1991. "How Should We Do the History of Statistics?" In *The Foucault Effect: Studies in Governmentality*, edited by Graham Burchell, Colin Gordon, and Peter Miller, 181–195. Chicago, IL: University of Chicago Press.

Hannerz, Ulf. 2016. *Writing Future Worlds: An Anthropologist Explores Global Scenarios.* London: Palgrave Macmillan.

Hazan, Haim, D. B. Bromley, Valerie Fennell, and Sharon Kaufman. 1984. "Continuity and Transformation among the Aged: A Study in the Anthropology of Time [and Comments]." *Current Anthropology* 25(5): 567–578.

Head, Brian W. 2008. "Wicked Problems in Public Policy." *Public Policy* 3(2): 101–118.

Heemskerk, Marieke. 2003. "Scenarios in Anthropology: Reflections on Possible Futures of the Suriname Maroons." *Futures* 35(9): 931–949.

Hermez, Sami Samir. 2017. *War Is Coming: Between Past and Future Violence in Lebanon.* Philadelphia, PA: University of Pennsylvania Press.

Hodges, Matt. 2008. "Rethinking Time's Arrow: Bergson, Deleuze and the Anthropology of Time." *Anthropological Theory* 8(4): 399–429.

Hoffmann, Markus, Hannah Kleine-Weber, Nadine Krüger, Marcel A. Mueller, Christian Drosten, and Stefan Pöhlmann. 2020. "The Novel Coronavirus 2019 (2019-nCoV) Uses the SARS-Coronavirus Receptor ACE2 and the Cellular Protease TMPRSS2 for Entry into Target Cells." *bioRxiv*, 31 January. https://doi.org/10.1101/2020.01.31.929042.

Holborn, Hajo. 1986. "The Prusso-German School: Moltke and the Rise of the General Staff." In *Makers of Modern Strategy from Machiavelli to the Nuclear Age*, edited by Peter Paret with the collaboration of Craig Gordon A. and Gilbert Felix, 281–295. Princeton, NJ: Princeton University Press.

Holton, Gerald. 1996. "On the Art of Scientific Imagination." *Daedalus* 125(2): 183–208.

Hounshell, David. 1997. "The Cold War, RAND, and the Generation of Knowledge, 1946–1962." *Historical Studies in the Physical and Biological Sciences* 27(2): 237–267.

Huang, Kristin. 2020. "World Health Organisation in Touch with Beijing after Mystery Viral Pneumonia Outbreak." *South China Morning Post*, 1 January. Available at: https://www.scmp.com/news/china/politics/article/3044207/china-shuts-seafood-market-linked-mystery-viral-pneumonia (accessed 22 March 2020).

Hui, David S., Esam I Azhar, Tariq A. Madani, Francine Ntoumi, Richard Kock, Osman Dar, Giuseppe Ippolito, et al. 2020. "The Continuing 2019-nCoV Epidemic Threat of Novel Coronaviruses to Global Health: The Latest 2019 Novel Coronavirus Outbreak in Wuhan, China." *International Journal of Infectious Diseases* 91: 264–266.

Huppauf, Bernd, and Christoph Wulf. 2013. "Introduction: The Indispensability of the Imagination." In *Dynamics and Performativity of Imagination: The Image between the Visible and the Invisible*, edited by Bernd Huppauf and Christoph Wulf, 1–18. New York: Routledge.

Institute for Health Metric and Evaluation (IHME). 2021. "COVID-19 Results Briefing: The European Union." *Institute for Health Metric and Evaluation*, 14 January. Available at: http://www.healthdata.org/sites/default/files/files/Projects/COVID/2021/briefing_EU_20210114.pdf (accessed 21 January 2021).

Jacob, François. 2001. "Imagination in Art and in Science." Translated by Tracy Ryan. *The Kenyon Review* 23(2): 113–121.

Jasanoff, Sheila. 1999. "The Songlines of Risk." *Environmental Values* 8(2): 135–152.

Jasanoff, Sheila. 2004. *States of Knowledge: The Co-Production of Science and the Social Order*. London: Routledge.

Jasanoff, Sheila. 2015. "Future Imperfect: Science, Technology, and the Imaginations of Modernity." In *Dreamscapes of Modernity: Sociotechnical Imaginaries and the Fabrication of Power*, edited by Sheila Jasanoff and Sang-Hyun Kim, 1–47. Chicago, IL: University of Chicago Press.

Jasanoff, Sheila, and Sang-Hyun Kim. 2013. "Sociotechnical Imaginaries and National Energy Policies." *Science as Culture* 22(2): 189–196.

Johns Hopkins Center for Health Security. 2020a. "Event 201: About." Johns Hopkins Center for Health Security, n.d. Available at: http://www.centerforhealthsecurity.org/event201/about (accessed 30 March 2020).

Johns Hopkins Center for Health Security. 2020b. "Event 201." Johns Hopkins Center for Health Security, n.d. Available at: http://www.centerforhealthsecurity.org/event201/ (accessed 30 March 2020).

Johns Hopkins Center for Health Security. 2020c. "Event 201: The Event 201 Scenario." Johns Hopkins Center for Health Security, n.d. Available at: http://www.centerforhealthsecurity.org/event201/scenario.html (accessed 30 March 2020).

Johns Hopkins Center for Health Security. 2020d. "Statement about nCoV and Our Pandemic Exercise." Johns Hopkins Center for Health Security, n.d. Available at:

http://www.centerforhealthsecurity.org/newsroom/center-news/2020-01-24
-Statement-of-Clarification-Event201.html (accessed 30 March 2020).

Kahane, Adam. 2012. *Transformative Scenario Planning: Working Together to Change the Future*. Oakland, CA: Berrett-Koehler.

Kahn, Herman. 1960. "The Arms Race and Some of Its Hazards." *Daedalus* 89(4): 744–780.

Kahn, Herman. 1965. "Herman Kahn Speaks at the Overseas Press Club, 16 June." New York Department of Records and Information Services. Available at: http://nycma .lunaimaging.com/luna/servlet/detail/RECORDSPHOTOUNITARC~26~26~13 34050~134909:MUNI-OSPC-1965-06-16-5785-7-T1425-T (accessed 20 February 2019).

Kahn, Herman, 1976. *"The Next Century": Margaret Mead, Herman Khan, William Irwin Thompson Discuss Nuclear Power*. Available at: https://youtu.be/1-QwGBDbd3M (accessed 24 Feb 2019).

Kahn, Herman. 1978. "Interview with Jack Webster." *Webster!*, 30 October. Available at: https://youtu.be/-3RYTMLJS70 (accessed 24 February 2019).

Kahn, Herman. 1984. *Thinking about the Unthinkable in the 1980s*. New York: Simon and Schuster.

Kahn, Herman. 2009a. "A Methodological Framework: The Alternative World Futures Approach." In *The Essential Herman Kahn: In Defense of Thinking*, edited by Paul Dragos Aligica and Kenneth R. Weinstein, 181–198. Plymouth, MA: Lexington Books.

Kahn, Herman. 2009b. "The Objectives of Future-Oriented Policy Research." In *The Essential Herman Kahn: In Defense of Thinking*, edited by Paul Dragos Aligica and Kenneth R. Weinstein, 155–166. Plymouth, MA: Lexington Books.

Kahn, Herman. 2009c. "Ways to Go Wrong." In *The Essential Herman Kahn: In Defense of Thinking*, edited by Paul Dragos Aligica and Kenneth R. Weinstein, 205–214. Plymouth, MA: Lexington Books.

Kahn, Herman. 2011. *On Thermonuclear War*. New Brunswick, NJ: Transaction.

Kahn, Herman, and Irwin Mann. 1957. *Techniques of Systems Analysis*. Santa Monica, CA: RAND Corporation. Available at: http://www.rand.org/pubs/research _memoranda/RM1829-1.html (accessed 9 September 2019).

Kahn, Herman, and Anthony J. Wiener. 1967. "The Next Thirty-Three Years: A Framework for Speculation." *Daedalus* 96(3): 705–732.

Kahneman, Daniel, and Amos Tversky. 1982. "Variants of Uncertainty." *Cognition* 11(2): 143–157.

Kama, Kärg. 2020. "Resource-Making Controversies: Knowledge, Anticipatory Politics and Economization of Unconventional Fossil Fuels." *Progress in Human Geography* 44(2): 333–356.

Kaufman, Sharon. 1994. "Old Age, Disease, and the Discourse on Risk: Geriatric Assessment in U.S. Health Care." *Medical Anthropology Quarterly* 8(4): 76–93.

Kaufmann, Mareile. 2016. "Exercising Emergencies: Resilience, Affect and Acting Out Security." *Security Dialogue* 47(2): 99–116.

Kearney, Richard. 1988. *The Wake of the Imagination: Toward a Postmodern Culture*. Minneapolis, MN: University of Minnesota Press.

Keck, Frédéric. 2018. "Avian Preparedness: Simulations of Bird Diseases and Reverse Scenarios of Extinction in Hong Kong, Taiwan, and Singapore." *Journal of the Royal Anthropological Institute* 24(2): 330–347.

Kelly, Mark G. E. 2010. *The Political Philosophy of Michel Foucault*. New York: Routledge.

Kemp, Adriana. 2000. "Border as Janus-Faced: Space and National Consciousness in Israel." *Theory and Criticism* 16: 13–44 [in Hebrew].

Khan, Natasha. 2020. "New Virus Discovered by Chinese Scientists Investigating Pneumonia Outbreak." *Wall Street Journal,* 8 January. Available at: https://www.wsj .com/articles/new-virus-discovered-by-chinese-scientists-investigating-pneumo nia-outbreak-11578485668 (accessed 22 March 2020).

Kimmerling, Baruch 1993. "Patterns of Militarism in Israel." *European Journal of Sociology* 34(2): 1–28.

Klar, Rebecca. 2020. "WHO Calls Emergency Meeting as Mystery Virus Spreads." *The Hill,* 20 January. Available at: https://thehill.com/policy/healthcare/479057-who -calls-emergency-meeting-as-mystery-virus-spreads (accessed 22 March 2020).

Kleiner, Art. 2003. "The Man Who Saw the Future." *Strategy + Business* 30. Available at: https://www.strategy-business.com/chapter/8220 (accessed 24 March 2017).

Knight, Frank H. 1921. *Risk, Uncertainty and Profit.* New York: Hart, Schaffner & Marx.

Kockelman, Paul, and Anya Bernstein. 2012. "Semiotic Technologies, Temporal Reckoning, and the Portability of Meaning. Or: Modern Modes of Temporality—Just How Abstract Are They?" *Anthropological Theory* 12(3): 320–348.

Koopman, Colin. 2013. *Genealogy as Critique: Foucault and the Problems of Modernity.* Bloomington, IN: Indiana University Press.

Koppenjan, Joop, and Erik-Hans Klijn. 2004. *Managing Uncertainties in Networks: A Network Approach to Problem Solving and Decision Making.* London: Routledge.

Krasmann, Susanne. 2015. "On the Boundaries of Knowledge: Security, the Sensible, and the Law." *InterDisciplines: Journal of History and Sociology* 6(2): 187–213.

Krasmann, Susanne, and Christine Hentschel. 2019. "'Situational Awareness': Rethinking Security in Times of Urban Terrorism." *Security Dialogue* 50(2): 181–197.

Krøijer, Stine. 2010. "Figurations of the Future: On the Form and Temporality of Protests among Left Radical Activists in Europe." *Social Analysis* 54(3): 139–152.

Kuchler, Magdalena. 2017. "Post-Conventional Energy Futures: Rendering Europe's Shale Gas Resources Governable." *Energy Research & Social Science* 31: 32–40.

Kuhn, Thomas S. 1962. *The Structure of Scientific Revolutions.* Chicago, IL: University of Chicago Press.

Labban, Mazen. 2011. "The Geopolitics of Energy Security and the War on Terror: The Case for Market Expansion and the Militarization of Global Space." In *Global Political Ecology,* edited by Richard Peet, Paul Robbins, and Michael Watts, 325–344. London: Routledge.

Lakatos, Imre. 1980. *The Methodology of Scientific Research Programmes.* Vol. 1, *Philosophical Papers.* Cambridge: Cambridge University Press.

Lakoff, Andrew. 2007. "Preparing for the Next Emergency." *Public Culture* 19(2): 247–271.

Lakoff, Andrew. 2008. "The Generic Biothreat, or, How We Became Unprepared." *Cultural Anthropology* 23(3): 399–428.

Lakoff, Andrew. 2015. "Global Health Security and the Pathogenic Imaginary." In *Dreamscapes of Modernity: Sociotechnical Imaginaries and the Fabrication of Power,* edited by Sheila Jasanoff and Sang-Hyun Kim, 301–320. Chicago, IL: University of Chicago Press.

Lakoff, Andrew. 2017. *Unprepared: Global Health in a Time of Emergency.* Berkeley, CA: University of California Press.

Lakoff, Andrew, and Stephen J. Collier, eds. 2008. *Biosecurity Interventions: Global Health and Security in Question.* New York: Columbia University Press.

Lentzos, Filippa, and Nikolas Rose. 2009. "Governing Insecurity: Contingency Planning, Protection, Resilience." *Economy and Society* 38(2): 230–254.

Lindley, Dennis V. 2000. "The Philosophy of Statistics." *Journal of the Royal Statistical Society: Series D (The Statistician)* 49(3): 293–337.

Lobo-Guerrero, Luis. 2011. *Insuring Security: Biopolitics, Security and Risk.* Abingdon: Routledge.

Lomsky-Feder, Edna, and Eyal Ben-Ari. 2000. *The Military and Militarism in Israeli Society.* Albany, NY: State University of New York Press.

Luhmann, Niklas. 1993. *Risk: A Sociological Theory.* New York: Aldine.

Luhmann, Niklas. 1998. *Observations on Modernity.* Stanford, CA: Stanford University Press.

Lupton, Deborah. 1994. *Moral Threats and Dangerous Desires: AIDS in the News Media.* London: Taylor & Francis.

Lupton, Deborah. 1999. *Risk and Sociocultural Theory: New Directions and Perspectives.* Cambridge: Cambridge University Press.

Margócsy, Dániel. 2017. "A Long History of Breakdowns: A Historiographical Review." *Social Studies of Science* 47(3): 307–325.

Masco, Joseph. 2006. *The Nuclear Borderlands.* Princeton, NJ: Princeton University Press.

Masco, Joseph. 2008. "'Survival Is Your Business': Engineering Ruins and Affect in Nuclear America." *Cultural Anthropology* 23(2): 361–398.

Masco, Joseph. 2014. *The Theater of Operations: National Security Affect from the Cold War to the War on Terror.* Durham, NC: Duke University Press.

Mason, Arthur. 2007. "The Rise of Consultant Forecasting in Liberalized Natural Gas Markets." *Public Culture* 19(2): 367–379.

Mathews, Andrew S., and Jessica Barnes. 2016. "Prognosis: Visions of Environmental Futures." *Journal of the Royal Anthropological Institute* 22(S1): 9–26.

Mey, Tim. 2006. "Imagination's Grip on Science." *Metaphilosophy* 37(2): 222–239.

Miyazaki, Hirokazu. 2006. *The Method of Hope: Anthropology, Philosophy, and Fijian Knowledge.* Stanford, CA: Stanford University Press.

Miyazaki, Hirokazu, and Annelise Riles. 2005. "Failure as an Endpoint." In *Global Assemblages: Technology, Politics, and Ethics as Anthropological Problems,* edited by Aihwa Ong and Stephen J. Collier, 320–331. Malden, MA: Blackwell.

Moats, Jason B., Thomas J. Chermack, and Larry M. Dooley. 2008. "Using Scenarios to Develop Crisis Managers: Applications of Scenario Planning and Scenario-Based Training." *Advances in Developing Human Resources* 10(3): 397–424.

Munn, Nancy D. 1992. "The Cultural Anthropology of Time: A Critical Essay." *Annual Review of Anthropology* 21: 93–123.

Nielsen, Morten. 2014. "A Wedge of Time: Futures in the Present and Presents without Futures in Maputo, Mozambique." *Journal of the Royal Anthropological Institute* 20(S1): 166–182.

O'Brien, Susie. 2016. "'We Thought the World Was Makeable': Scenario Planning and Postcolonial Fiction." *Globalizations* 13(3): 329–344.

Ogilvy, James A. 2002. *Creating Better Futures: Scenario Planning as a Tool for a Better Tomorrow.* Oxford: Oxford University Press.

Ogilvy, Jay, and Erik Smith. 2004. "Mapping Public and Private Scenario Planning: Lessons from Regional Projects." *Development* 47(4): 67–72.

O'Malley, Pat. 1992. "Risk, Power and Crime Prevention." *Economy and Society* 21(3): 252–275.

O'Malley, Pat. 2004. *Risk, Uncertainty and Government.* London: Glasshouse.

O'Malley, Pat. 2010. "Resilient Subjects: Uncertainty, Warfare and Liberalism." *Economy and Society* 39(4): 488–509.

O'Malley, Pat. 2015. "Uncertainty Makes Us Free: Insurance and Liberal Rationality." In *Modes of Uncertainty: Anthropological Cases,* edited by Limor Samimian-Darash and Paul Rabinow, 13–28. Chicago, IL: University of Chicago Press.

Opitz, Sven. 2017. "Simulating the World: The Digital Enactment of Pandemics as a Mode of Global Self-Observation." *European Journal of Social Theory* 20(3): 392–416.

Opitz, Sven, and Ute Tellmann. 2015. "Future Emergencies: Temporal Politics in Law and Economy." *Theory, Culture and Society* 32(2): 107–129.

Osborne, David. 1993. "Reinventing Government." *Public Productivity and Management Review* 16(4): 349–356.

Pan American Health Organization (PAHO). 2020. "PAHO Director Urges Readiness to Detect Cases of New Coronavirus in the Americas." *Pan American Health Organization*, 24 January. Available at: https://www.paho.org/hq/index.php?option =com_content&view=article&id=15694:paho-director-urges-readiness-to-det ect-cases-of-new-coronavirus-in-the-americas&Itemid=1926&lang=en (accessed 29 March 2020).

Pappas, Nickolas. 2004. *Routledge Philosophy Guidebook to Plato and the* Republic. London: Routledge.

Paret, Peter. 2007. *Clausewitz and the State: The Man, His Theories, and His Times*. Princeton, NJ: Princeton University Press.

Peebles, Gustav. 2009. "Inverting the Panopticon: Money and the Nationalization of the Future." *Public Culture* 20(2): 233–265.

Petryna, Adriana. 2015. "What Is a Horizon? Navigating Thresholds in Climate Change Uncertainty." In *Modes of Uncertainty: Anthropological Cases*, edited by Limor Samimian-Darash and Paul Rabinow, 147–164. Chicago, IL: University of Chicago Press.

Pias, Claus. 2011. "On the Epistemology of Computer Simulation." *Zeitschrift Für Medien-Und Kulturforschung* 1: 29–54.

Popper, Karl. 1963. *Conjectures and Refutations: The Growth of Scientific Knowledge*. New York: Routledge.

Porter, Theodore M. 1996. *Trust in Numbers: The Pursuit of Objectivity in Science and Public Life*. Princeton, NJ: Princeton University Press.

Power, Michael. 2004. *The Risk Management of Everything: Rethinking the Politics of Uncertainty*. New York: Demos.

ProMED-mail. 2019. "Undiagnosed Pneumonia—China (Hubei): Request for Information." The International Society for Infectious Diseases, 30 December. Available at: https://promedmail.org/promed-post/?id=20191230.6864153 (accessed 22 March 2020).

ProMED-mail. 2020a. "Undiagnosed Pneumonia—China (Hubei) (02): Updates, Other Country Responses, Request for Information." The International Society for Infectious Diseases, 3 January. Available at: https://promedmail.org/promed-post /?id=20200103.6869668 (accessed 22 March 2020).

ProMED-mail. 2020b. "Undiagnosed Pneumonia—China (Hubei) (02): Undiagnosed Pneumonia—China (HU) (05): Novel Coronavirus Identified." The International Society for Infectious Diseases, 8 January. Available at: https://promedmail.org /promed-post/?id=20200108.6877694 (accessed 22 March 2020).

ProMED-mail. 2020c. "Novel Coronavirus (12): China (HU) New Fatality, Healthcare Workers, WHO." The International Society for Infectious Diseases, 21 January. Available at: https://promedmail.org/promed-post/?id=20200121.6901757 (accessed 22 March 2020).

ProMED-mail. 2020d. "Novel Coronavirus (16): China (HU) Viet Nam, Singapore ex China, IHR Committee WHO." The International Society for Infectious Diseases, 23 January. Available at: https://promedmail.org/promed-post/?id=20200123.6910685 (accessed 22 March 2020).

ProMED-mail. 2020e. "COVID-19 Update (27): Global, Egypt, USA, Taiwan, More Countries, WHO." The International Society for Infectious Diseases, 4 March. Available at: https://promedmail.org/promed-post/?id=20200304.7046372 (accessed 22 March 2020).

ProMED-mail. 2020f. "COVID-19 Update (487): Denmark, Animal, Mink, Zoonotic, Risk Assessment ECDC." The International Society for Infectious Diseases, 12 November. Available at: https://promedmail.org/promed-post/?id=7939110 (accessed 20 January 2021).

ProMED-mail. 2020g. "COVID-19 Update (539): UK New Variant, Genomic Study on Severity, WHO, Global." The International Society for Infectious Diseases, 15 December. Available at: https://promedmail.org/promed-post/?id=8018909 (accessed 20 January 2021).

ProMED-mail. 2020h. "COVID-19 Update (545): Mutations, UK, South Africa, South Asia, WHO, Global." The International Society for Infectious Diseases, 20 December. Available at: https://promedmail.org/promed-post/?id=8032607 (accessed 20 January 2021).

ProMED-mail. 2020i. "COVID-19 Update (553): Variants, Immunocompromised, M East/N Africa, WHO, Global." The International Society for Infectious Diseases, 24 December. Available at: https://promedmail.org/promed-post/?id=8043651 (accessed 20 January 2021).

ProMED-mail. 2021a. "COVID-19 Update (16): Vaccines, Tracking Variants, USA, Travel, WHO, Global." The International Society for Infectious Diseases, 14 January. Available at: https://promedmail.org/promed-post/?id=8102385 (accessed 20 January 2021).

ProMED-mail. 2021b. "COVID-19 Update (20): Animal, Deer, Experimental Infection." The International Society for Infectious Diseases, 16 January. Available at: https://promedmail.org/promed-post/?id=8108967 (accessed 20 January 2021).

ProMED-mail. 2021c. "COVID-19 Update (24): Vaccine, New Variant, WHO, Global." The International Society for Infectious Diseases, 19 January. Available at: https://promedmail.org/promed-post/?id=8115933 (accessed 20 January 2021).

Rabinow, Paul. 2003. *Anthropos Today*. Princeton, NJ: Princeton University Press.

Rabinow, Paul. 2007. *Marking Time: On the Anthropology of the Contemporary*. Princeton, NJ: Princeton University Press.

Rabinow, Paul, and Gaymon Bennett. 2012. *Designing Human Practices: An Experiment with Synthetic Biology*. Chicago, IL: University of Chicago Press.

Rabinow, Paul, and Nikolas Rose. 2003. "Introduction: Foucault Today." In *The Essential Foucault: Selections from Essential Works of Foucault, 1954–1984*, edited by Paul Rabinow and Nikolas Rose, vii–xxxv. New York: New Press.

Ramírez, Rafael, and Angela Wilkinson. 2016. *Strategic Reframing: The Oxford Scenario Planning Approach*. Oxford: Oxford University Press.

Rapp, Rayna. 1995. "Risky Business: Genetic Counseling in a Shifting World." In *Articulating Hidden Histories: Exploring the Influence of Eric R. Wolf*, edited by Jane Schneider and Rayna Rapp, 175–189. Berkeley, CA: University of California Press.

Ratcliffe, John. 2000. "Scenario Building: A Suitable Method for Strategic Property Planning?" *Property Management* 18(2): 127–144.

Ravetz, Jerome R. 1999. "What Is Post-Normal Science?" *Futures* 31: 647–653.

Ravetz, Jerome R. 2006. "Post-Normal Science and the Complexity of Transitions towards Sustainability." *Ecological Complexity* 3(4): 275–284.

Reid, Julian. 2012. "The Disastrous and Politically Debased Subject of Resilience." *Development Dialogue* 58(1): 67–79.

Revet, Sandrine. 2013. "'A Small World': Ethnography of a Natural Disaster Simulation in Lima, Peru." *Social Anthropology* 21(1): 38–53.

Richardson, Alan. 2011. "Defaulting to Fiction: Neuroscience Rediscovers the Romantic Imagination." *Poetics Today* 32(4): 663–692.

Richardson, Alan. 2013. "Reimagining the Romantic Imagination." *European Romantic Review* 24(4): 385–402.

Ringel, Felix. 2016. "Beyond Temporality: Notes on the Anthropology of Time from a Shrinking Fieldsite." *Anthropological Theory* 16(4): 390–412.

Ringel-Hoffman, Ariella. 2015. "Apocalypse Tomorrow." *Yedioth Ahronoth*, 29 May.

Ringland, Gill. 1998. *Scenario Planning: Managing for the Future*. Chichester: John Wiley & Sons.

Rittel, Horst W., and Melvin M. Webber. 1973. "Planning Problems Are Wicked." *Polity* 4(155): 155–169.

Robbins, Joel. 2010. "On Imagination and Creation: An Afterword." *Anthropological Forum* 20(3): 305–313.

Rose, Nikolas. 1996. "Psychiatry as a Political Science: Advanced Liberalism and the Administration of Risk." *History of the Human Sciences* 9(2): 1–23.

Rose, Nikolas. 2007. *The Politics of Life Itself: Biomedicine, Power, and Subjectivity in the Twenty-First Century*. Princeton, NJ: Princeton University Press.

Rosenberg, Daniel, and Susan Harding. 2005. "Introduction: Histories of the Future." In *Histories of the Future*, edited by Daniel Rosenberg and Susan Harding, 1–18. Durham, NC: Duke University Press.

Rosten, Leo. 1941. *Hollywood: The Movie Colony, the Movie Makers*. Riverdale, NY: Ayer.

Rothenberg, Gunther E. 1996. "Moltke, Schlieffen, and the Doctrine of Strategic Envelopment." In *Makers of Modern Strategy from Machiavelli to the Nuclear Age*, edited by Paret Peter, with the collaboration of Craig Gordon A. and Gilbert Felix, 296–325. Princeton, NJ: Princeton University Press.

Rubin, Rita, Jennifer Abbasi, and Rebecca Voelker. 2020. "Latin America and Its Global Partners Toil to Procure Medical Supplies as COVID-19 Pushes the Region to Its Limit." *JAMA* 324(3): 217–219. doi:10.1001/jama.2020.11182.

Runde, Jochen. 1998. "Clarifying Frank Knight's Discussion of the Meaning of Risk and Uncertainty." *Cambridge Journal of Economics* 22(5): 539–546.

Salis, Fiora, and Roman Frigg. Forthcoming. "Capturing the Scientific Imagination." In *The Scientific Imagination*, edited by Peter Godfrey-Smith and Arnon Levy. Oxford: Oxford University Press.

Samimian-Darash, Limor. 2009. "A Pre-Event Configuration for Biological Threats: Preparedness and the Constitution of Biosecurity Events." *American Ethnologist* 36(3): 478–491.

Samimian-Darash, Limor. 2011a. "Governing through Time: Preparing for Future Threats to Health and Security." *Sociology of Health and Illness* 33(6): 930–945.

Samimian-Darash, Limor. 2011b. "The Re-Forming State: Actions and Repercussions in Preparing for Future Biological Events." *Anthropological Theory* 11(3): 283–307.

Samimian-Darash, Limor. 2013. "Governing Future Potential Biothreats: Toward an Anthropology of Uncertainty." *Current Anthropology* 54(1): 1–22.

Samimian-Darash, Limor. 2016. "Practicing Uncertainty: Scenario-Based Preparedness Exercises in Israel." *Cultural Anthropology* 31(3): 359–386.

Samimian-Darash, Limor, and Paul Rabinow, eds. 2015. *Modes of Uncertainty: Anthropological Cases*. Chicago, IL: University of Chicago Press.

Samimian-Darash, Limor, and Nir Rotem. 2019. "From Crisis to Emergency: The Shifting Logic of Preparedness." *Ethnos* 84(5): 910–926.

Schauble, Michaela. 2016. "Introduction: Mining Imagination—Ethnographic Approaches beyond the Written Word." *Anthrovision* 4(2): 1–10.

Schoch-Spana, Monica. 2004. "Bioterrorism: US Public Health and a Secular Apocalypse." *Anthropology Today* 20(5): 8–13.

Schwartz, Peter. 1996. *The Art of the Long View: Paths to Strategic Insight for Yourself and Your Company.* New York: Doubleday.

Schwartz, Peter. 2002. "The River and the Billiard Ball: History, Innovation, and the Future." In *What the Future Holds: Insights from Social Science*, edited by Richard N. Cooper and Richard Layard, 17–28. Cambridge, MA: MIT Press.

Seipel, Brooke. 2020. "FDA Creates First-Ever Medical Supply Shortage List Including Masks, Swabs and Ventilators." *The Hill*, 14 August. Available at: https://thehill .com/policy/healthcare/512129-fda-creates-first-ever-medical-supply-shortage -list-including-masks-swabs (accessed 20 January 2021).

Sha, Richard C. 2009. "Imagination as Inter-Science." *European Romantic Review* 20(5): 661–669.

Sha, Richard C. 2013. "Romantic Physiology and the Work of Romantic Imagination: Hypothesis and Speculation in Science and Coleridge." *European Romantic Review* 24(4): 403–419.

Shapiro, Matan, and Nurit Bird-David. 2017. "Routinergency: Domestic Securitization in Contemporary Israel." *Environment and Planning D: Society and Space* 35(4): 637–655.

Sheffer, Gabriel, and Oren Barak, eds. 2010 *Militarism and Israeli Society*. Bloomington, IN: Indiana University Press.

Shell International. 2003. *Scenarios: An Explorer's Guide.* London: Shell Business Environment, Shell International.

Shir-Vertesh, Dafna, and Fran Markowitz. 2015. "Between War and Peace: Israel Day by Day." *Ethnologie francaise* 45(2): 209–221.

Shklar, Judith. 1965. "The Political Theory of Utopia: From Melancholy to Nostalgia." *Daedalus* 94(2): 367–381.

Siddiqui, Faiz. 2020. "The U.S. Forced Major Manufacturers to Build Ventilators. Now They're Piling Up Unused in a Strategic Reserve." *Washington Post*, 18 August. Available at: https://www.washingtonpost.com/business/2020/08/18/ventilators -coronavirus-stockpile/?hpid=hp_hp-top-table-main_ventilators-715am%3Aho mepage%2Fstory-ans (accessed 20 January 2021).

Simon, Stephanie, and Marieke de Goede. 2015. "Cybersecurity, Bureaucratic Vitalism and European Emergency." *Theory, Culture and Society* 32(2): 79–106.

Sneath, David, Martin Holbraad, and Morten Axel Pedersen. 2009. "Technologies of the Imagination: An Introduction." *Ethnos* 74(1): 5–30.

Stephenson, Niamh, and Michelle Jamieson. 2009. "Securitising Health: Australian Newspaper Coverage of Pandemic Influenza." *Sociology of Health and Illness* 31(4): 525–539.

Stone, Will, and Sean McMinn. 2020. "Near Crisis, Some Hospitals Face Tough Decisions in Caring for Floods of Patients." *NPR*, 25 November. Available at: https://www.npr .org/sections/health-shots/2020/11/25/939103515/near-crisis-some-hospitals-face -tough-decisions-in-caring-for-floods-of-patients (accessed 20 January 2021).

Stuart, Michael T. 2017. "Imagination: A Sine Qua Non of Science." *Croatian Journal of Philosophy* 17(49): 9–32.

Tabak, John. 2014. *Probability and Statistics: The Science of Uncertainty.* New York: Infobase Publishing.

Taleb, Nassim Nicholas. 2007. *The Black Swan: The Impact of the Highly Improbable.* New York: Random House.

Taylor, Charles. 2004. *Modern Social Imaginaries*. Durham, NC: Duke University Press.

Tellmann, Ute. 2009. "Imagining Catastrophe: Scenario Planning and the Striving for Epistemic Security." *Economic Sociology: The European Electronic Newsletter* 10(2): 17–21.

Thomas, Nigel J. T. 1999. "Imagination." In *Dictionary of Philosophy of Mind*, edited by Chris Eliasmith and Eric Hochstein. Available at: http://dictionaryofphilmind .com/imagination (accessed 16 March 2018).

Tolon, Kaya. 2011. *The American Futures Studies Movement (1965–1975): Its Roots, Motivations, and Influences*. Doctoral dissertation, Iowa State University.

Volkery, Axel, and Teresa Ribeiro. 2009. "Scenario Planning in Public Policy: Understanding Use, Impacts and the Role of Institutional Context Factors." *Technological Forecasting and Social Change* 76(9): 1198–1207.

Wack, Pierre. 1985a. "Scenarios: Uncharted Waters Ahead." *Harvard Business Review* 63(5): 72–89.

Wack, Pierre. 1985b. "Shooting the Rapids." *Harvard Business Review* 63(6): 139–150.

Wack, Pierre. 1986. "Scenario Planning: Planning in Turbulent Times." *Oxford Futures Library*. Available at: http://oxfordfutures.sbs.ox.ac.uk/pierre-wack-memorial -library/video/index.html (accessed 24 February 2019).

Wallman, Sandra. 2003. "Introduction: Contemporary Futures." In *Contemporary Futures: Perspectives from Social Anthropology*, edited by Sandra Wallman, 1–20. London: Routledge.

Wardle, Huon. 2015. "Afterword: An End to Imagining?" In *Reflections on Imagination: Human Capacity and Ethnographic Method*, edited by Mark Harris and Nigal Rapport, 275–294. Farnham: Ashgate.

Weaver, Warren. 1929. "Science and Imagination." *The Scientific Monthly* 29(5): 425–434.

Weber, Max. [1917] 2009. "Science as a Vocation." In *From Max Weber: Essays in Sociology*, edited by Hans Heinrich Gerth and C. Wright Mills, 129–159. New York: Routledge.

Weszkalnys, Gisa. 2014. "Anticipating Oil: The Temporal Politics of a Disaster Yet to Come." *The Sociological Review* 62: 211–235.

Weszkalnys, Gisa. 2015. "Geology, Potentiality, Speculation: On the Indeterminacy of First Oil." *Cultural Anthropology* 30(4): 611–639.

Wilkinson, Angela, and Roland Kupers. 2013. "Living in the Futures." *Harvard Business Review* 91(5): 118–127.

Wilson, Ian. 2000. "From Scenario Thinking to Strategic Action." *Technological Forecasting and Social Change* 65(1): 23–29.

Winsberg, Eric B. 2010. *Science in the Age of Computer Simulation*. Chicago, IL: University of Chicago Press.

Wodak, Josh, and Timothy Neale. 2015. "A Critical Review of the Application of Environmental Scenario Exercises." *Futures* 73: 176–186.

World Energy Council. 2016. *World Energy Scenarios 2016: The Grand Transition*. London: World Energy Council.

World Energy Council. 2018. *World Energy Council "Grand Transition" Scenarios: Focus on the European Region* [Draft]. London: World Energy Council.

World Energy Council. 2019. *World Energy Insights Brief: 2019; Technical Annex*. London: World Energy Council.

World Health Organization (WHO). 2005. *International Health Regulations*. 3rd ed. Geneva: WHO.

World Health Organization (WHO). n.d. "Exercise Management Team Terms of Reference Guidance Note. Functional Exercise Document FX04." Available through the

WHO Strategic Partnership for IHR and Health Security platform at: https://extranet.who.int/sph/simulation-exercise (accessed 25 September 2019).

World Health Organization (WHO). 2020a. "Coronavirus Disease 2019 (COVID-19) Situation Report—43." World Health Organization, 3 March. Available at: https://www.who.int/docs/default-source/coronaviruse/situation-reports/20200303-sitrep-43-covid-19.pdf?sfvrsn=76e425ed_2 (accessed 22 March 2020).

World Health Organization (WHO). 2020b. "Coronavirus Disease 2019 (COVID-19) Situation Report—89." World Health Organization, 18 April. Available at: https://www.who.int/docs/default-source/coronaviruse/situation-reports/20200418-sitrep-89-covid-19.pdf?sfvrsn=3643dd38_2 (accessed 19 April 2020).

World Health Organization (WHO). 2020c. "WHO Director-General's Opening Remarks at the Media Briefing on COVID-19." World Health Organization, 11 March. Available at: https://www.who.int/dg/speeches/detail/who-director-general-s-opening-remarks-at-the-media-briefing-on-covid-19–11-march-2020 (accessed 27 March 2020).

World Health Organization (WHO). 2020d. "Novel Coronavirus—China. Disease Outbreak News: Update." World Health Organization, 12 January. Available at: https://www.who.int/csr/don/12-january-2020-novel-coronavirus-china/en/ (accessed 27 March 2020).

World Health Organization (WHO). 2020e. "Updated WHO Advice for International Traffic in Relation to the Outbreak of the Novel Coronavirus 2019-nCoV." World Health Organization, 24 January. Available at: https://www.who.int/news-room/articles-detail/updated-who-advice-for-international-traffic-in-relation-to-the-outbreak-of-the-novel-coronavirus-2019-ncov-24-jan/ (accessed 27 March 2020).

World Health Organization (WHO). 2020f. "Novel Coronavirus (2019-nCoV) Situation Report—4." World Health Organization, 24 January. Available at: https://www.who.int/docs/default-source/coronaviruse/situation-reports/20200124-sitrep-4-2019-ncov.pdf?sfvrsn=9272d086_8 (accessed 27 March 2020).

World Health Organization (WHO). 2020g. "Novel Coronavirus (2019-nCoV) Situation Report—2." World Health Organization, 21 January. Available at: https://www.who.int/docs/default-source/coronaviruse/situation-reports/20200122-sitrep-2-2019-ncov.pdf?sfvrsn=4d5bcbca_2 (accessed 29 March 2020).

World Health Organization (WHO). 2020h. "Statement on the Meeting of the International Health Regulations (2005) Emergency Committee Regarding the Outbreak of Novel Coronavirus (2019-nCoV)." World Health Organization, 23 January. Available at: https://www.who.int/news-room/detail/23-01-2020-statement-on-the-meeting-of-the-international-health-regulations-(2005)-emergency-committee-regarding-the-outbreak-of-novel-coronavirus-(2019-ncov) (accessed 29 March 2020).

World Health Organization (WHO). 2020i. "WHO Director-General's Statement on the Advice of the IHR Emergency Committee on Novel Coronavirus." World Health Organization, 23 January. Available at: https://www.who.int/dg/speeches/detail/who-director-general-s-statement-on-the-advice-of-the-ihr-emergency-committee-on-novel-coronavirus (accessed 29 March 2020).

World Health Organization (WHO). 2020j. "Statement on the Second Meeting of the International Health Regulations (2005) Emergency Committee Regarding the Outbreak of Novel Coronavirus (2019-nCoV)." World Health Organization, 30 January. Available at: https://www.who.int/news-room/detail/30-01-2020-statement-on-the-second-meeting-of-the-international-health-regulations-(2005)-emergency-committee-regarding-the-outbreak-of-novel-coronavirus-(2019-ncov) (accessed 29 March 2020).

World Health Organization (WHO). 2020k. "Coronavirus Disease 2019 (COVID-19) Situation Report—39." World Health Organization, 28 February. Available at: https://

www.who.int/docs/default-source/coronaviruse/situation-reports/20200228
-sitrep-39-covid-19.pdf?sfvrsn=5bbf3e7d_4 (accessed 29 March 2020).

World Health Organization (WHO). 2020l. "WHO Director-General's Opening Remarks at the Media Briefing on COVID-19." World Health Organization, 28 February. Available at: https://www.who.int/dg/speeches/detail/who-director-general-s-open ing-remarks-at-the-media-briefing-on-covid-19–28-february-2020 (accessed 29 March 2020).

World Health Organization (WHO). 2020m. "WHO Director-General's Opening Re-marks at the Media Briefing on COVID-19." World Health Organization, 27 Feb-ruary. Available at: https://www.who.int/dg/speeches/detail/who-director-general -s-opening-remarks-at-the-media-briefing-on-covid-19–27-february-2020 (accessed 29 March 2020).

World Health Organization (WHO). 2020n. "Coronavirus Disease (COVID-19) Press Conference." World Health Organization, 28 February. Available at: https://www .who.int/docs/default-source/coronaviruse/transcripts/who-audio-emergencies -coronavirus-press-conference-full-28feb2020.pdf?sfvrsn=13eeb6a4_2 (accessed 29 March 2020).

World Health Organization (WHO). 2020o. "Coronavirus Disease (COVID-19) Train-ing: Simulation Exercise." World Health Organization, n.d. Available at: https:// www.who.int/emergencies/diseases/novel-coronavirus-2019/training/simulation -exercise (accessed 30 March 2020).

World Health Organization (WHO). 2020p. "Novel Coronavirus (COVID-19): Health Emergency Preparedness Simulation Exercise. PowerPoint Presentation." World Health Organization, n.d. Available at: https://www.who.int/emergencies/diseases /novel-coronavirus-2019/training/simulation-exercise (accessed 30 March 2020).

World Health Organization (WHO). 2020q. "Namibia COVID19 TTX Mar.2020." World Health Organization, 17 March. Available at: https://extranet.who.int/sph/namibia -covid19-ttx-mar2020 (accessed 30 March 2020).

World Health Organization (WHO). 2020r. "Ethiopia COVID19 SimEx." World Health Organization, 5–6 March. Available at: https://extranet.who.int/sph/ethiopia-covid 19-simex (accessed 30 March 2020).

World Health Organization (WHO). 2020s. "Coronavirus Disease 2019 (COVID-19) Sit-uation Report—99." World Health Organization, 28 April. Available at: https:// www.who.int/docs/default-source/coronaviruse/situation-reports/20200428 -sitrep-99-covid-19.pdf (accessed 20 January 2021).

World Health Organization (WHO). 2020t. "COVID-19 Essential Supplies Forecasting Tool." World Health Organization, 26 August. Available at: https://www.who.int /publications/m/item/covid-19-essential-supplies-forecasting-tool (accessed 21 Jan-uary 2021).

World Health Organization (WHO). 2021. "Weekly Epidemiological Update." World Health Organization, 19 January. Available at: https://www.who.int/publications/m /item/weekly-epidemiological-update—19-january-2021 (accessed 20 January 2021).

World Health Organization European Regional Office (WHO EURO). 2019a. *Exercise JADE. Report, 2018.* Copenhagen: WHO Regional Office for Europe.

World Health Organization European Regional Office (WHO EURO). 2019b. *Exercise JADE: Injects.* Copenhagen: WHO.

World Health Organization European Regional Office (WHO EURO). 2020a. "Adaptt Surge Planning Support Tool." World Health Organization, n.d. Available at: https://euro.sharefile.com/share/view/scef08a92a9d43b68/fo62fb4f-ba90-4a9a -a0e7-7c98dea86b9a (accessed 21 January 2021).

World Health Organization European Regional Office (WHO EURO). 2020b. "Health Workforce Estimator (HWFE)." World Health Organization, n.d. Available at: https://euro.sharefile.com/share/view/s1df028894aa49abb/fob92ed8-23cb-4b24 -a746-524bb6a27843 (accessed 21 January 2021).

World Health Organization Regional Office for Africa (WHO AFRO). 2020. "Easing COVID-19 Impact on Key Health Services." World Health Organization, 5 November. Available at: https://www.afro.who.int/news/easing-covid-19-impact-key -health-services (accessed 20 January 2021).

World Health Organization Western Pacific Regional Office. 2011. *IHR Exercise Crystal.* Manila: World Health Organization. Available at: https://apps.who.int/iris/bitstream /handle/10665/208557/WPDSE1206256_ESR_eng.pdf?sequence=1&isAllowed=y (accessed August 2019).

World Health Organization Western Pacific Regional Office. 2015. *IHR Exercise Crystal.* Manila: World Health Organization. Available at: https://apps.who.int/iris /bitstream/handle/10665/246428/20151204-PHL-eng.pdf?sequence=1&isAllowed =y (accessed August 2019).

Wright, George, and Paul Goodwin. 2009. "Decision Making and Planning Under Low Levels of Predictability: Enhancing the Scenario Method." *International Journal of Forecasting* 25(4): 813–825.

Wright, George, Kees van der Heijden, George Burt, Ron Bradfield, and George Cairns. 2008. "Scenario Planning Interventions in Organizations: An Analysis of the Causes of Success and Failure." *Futures* 40(3): 218–236.

Wu, Aiping, Yousong Peng, Baoying Huang, Xiao Ding, Xianyue Wang, Peihua Niu, Jing Meng, et al. 2020. "Genome Composition and Divergence of the Novel Coronavirus (2019-nCoV) Originating in China." *Cell Host and Microbe* 27: 325–328.

Wynne, Brian. 2002. "Risk and Environment as Legitimatory Discourses of Technology: Reflexivity Inside Out?" *Current Sociology* 50(3): 459–477.

Zalik, Anna. 2010. "Oil 'Futures': Shell's Scenarios and the Social Constitution of the Global Oil Market." *Geoforum* 41(4): 553–564.

Zeiderman, Austin. 2016. *Endangered City: The Politics of Security and Risk in Bogotá.* Durham, NC: Duke University Press.

Zeitlyn, David. 2015. "Looking Forward, Looking Back." *History and Anthropology* 26(4): 381–407.

Zimmer, Carl. 2021. "New California Variant May Be Driving Virus Surge There, Study Suggests." *New York Times*, 19 January. Available at: https://www.nytimes.com /2021/01/19/health/coronavirus-variant-california.html?action=click&module =Well&pgtype=Homepage%C2%A7ion=Health (accessed 21 January 2021).

Zinn, Jen O. 2009. "The Sociology of Risk and Uncertainty: A Response to Judith Green's 'Is It Time for the Sociology of Health to Abandon "Risk"?'" *Health, Risk and Society* 11(6): 509–526.

Index

actualizations, 11, 60–66, 71, 136, 147, 160n12, 160n13. *See also* counter-actualizations

anthropology and scenario technology, 4–7, 160n6. *See also* scenario technology

anticipation, 5, 15, 16, 17, 162n1 (ch. 2), 162n4

anticipatory governance, 13

business industry and scenarios, 32, 73. *See also* scenario technology

Center for Health Security (Johns Hopkins University), 143–45

climate change, 4, 17, 74, 75, 78

Cold War, 27–28

Coronavirus. *See* COVID-19 pandemic

counter-actualizations, 66–71. *See also* actualizations

counter-effectuated, 71. *See also* actualizations

"COVID-19 Essential Supplies Forecasting Tool (ESFT)" (WHO), 134–35

COVID-19 pandemic, 6, 133–46, 163n1 (ch. 7). *See also* disease outbreaks

Crystal exercises, 100–101, 106–7, 112, 113, 122. *See also* JADE (Joint Assessment and Detection of Events) exercises; scenario-based exercises

danger *vs.* risk, 9. *See also* risk

Delphi technique, 29, 39

disease outbreaks: COVID-19, 6, 133–46; Ebola, 100; H1N1, 100; H7N7, 100; HIV/AIDS, 163n1 (ch. 5); influenza, ix, 98, 139, 140, 159n1; *Listeria,* 97, 101, 106, 111, 115, 123; MERS, 163n1 (ch. 7); SARS, 122, 133, 135, 137, 139, 144, 163n1 (ch. 5), 163n1 (ch. 6), 163n5; scenario exercise of, 143–46; smallpox, ix, 159n1, 163n5; vaccinations and, ix, 135–36, 144, 159n1. *See also* public health emergency scenarios

earthquake scenarios, 5, 44–45, 56, 57, 65–66. *See also* scenario-based exercises

Ebola outbreaks, 100. *See also* disease outbreaks

effectuated, 71. *See also* actualizations

EIS (Event Information Site), 101, 111, 114, 115, 163n4

Emergency Risk Communication (ERC) plan, 108

energy scenarios, 32, 73–76, 126–29, 151. *See also* scenario-based exercises; World Energy Council

Enlightenment, 148, 149–50

evacuation drills, 62–63

Event 201 exercise, 143–45. *See also* scenario-based exercises

Event Information Site (EIS), 101, 111, 114, 115, 163n4

food safety scenarios, 97, 101, 106, 111, 115, 123. *See also* disease outbreaks; scenario-based exercises

future, 14–16, 71, 125–29, 146–47. *See also* scenario technology; temporality; time

future presents, 117

governmentality, 11–14, 18, 95, 161n15. *See also* security

H1N1 pandemic, 100. *See also* disease outbreaks

H7N7 outbreak, 100. *See also* disease outbreaks

"Health Workforce Estimator" (WHO), 135

HIV/AIDS, 163n1 (ch. 5). *See also* disease outbreaks

IHR (international health regulations), 99–100, 105–15, 163nn1–3 (ch. 5). *See also* Crystal exercises; JADE (Joint Assessment and Detection of Events) exercises

imagination: scenarios and, 150–53; science and, 148–50, 153–55, 164n7; uncertainty by design and, 2

indetermination, 16–17, 20, 55, 60–66, 129

influenza, ix, 98, 139, 140, 159n1. *See also* disease outbreaks

insurance, 4, 12, 13, 17

international health regulations. *See* IHR (international health regulations)

CPSIA information can be obtained
at www.ICGtesting.com
Printed in the USA
LVHW041934110322
713099LV00004B/489

9 781501 762468